AF709823

BULTACO

Limited Edition Extra

1971-1979

Compiled by R M Clarke

ISBN 1 85520 6927

BROOKLANDS BOOKS LTD.
P.O. BOX 146, COBHAM,
SURREY, KT11 1LG. UK
sales@brooklands-books.com

www.brooklands-books.com

A-BU7BX2

Printed in China

Brooklands Books

ACKNOWLEDGEMENTS

Every book starts somewhere, and the idea for this one started whilst holidaying in Madrid in the autumn of 2004. Whilst there I took the opportunity of calling in on our long time customer, Pablo Villanuera, who runs the splendid Libro Motor bookshop. Unfortunately, Pablo and I have something in common - a lack of linguistic skills. One point however came across clearly: why had we never produced a book on a Spanish car or motorcycle? The outcome is two books on Bultaco, this the first, covering the sixties and its companion volume covering the seventies. Pablo helped us in so many ways - finding photographs, giving advice and much more for which we are grateful.

Our thanks also go to our talented friend and agent Joe Portelli who used his translation skills to bring these books to fruition. In addition we would like to thank the publishers of *Motorcyclist, Cycle World, Modern Cycle, Motorcycle Mechanics, Cycle Guide, Popular Cycling* and *Motorcycle Enthusiast* for allowing us once again to include their interesting copyright articles in this reference portfolio.

R M Clarke

Having made it successfully through the first years of its birth, and then succeeded internationally in the 60s, Bultaco expanded its model range and continued with its "race on Sunday sell on Monday" policy.

Racing in this case was in the tough enduro off-road field where rugged construction and engine flexibility was as important as rider skill. In fact one of the enduring features of Bultaco engines was their amazing balance between power and torque, ideal for the type of competition at which the factory was so adept having specialised in that field. With all of its international success, the Barcelona-based company was really a minnow among the world's motorcycle manufacturers. Where the Japanese factories of Honda, Yamaha and Suzuki were churning out millions of bikes each year, Bultaco managed just 15,000 in 1971. Even at that modest figure, it was the largest of the Spanish motorcycle manufacturers.

Comments from the media were generally positive, especially where the Bultaco's engine performance, braking and handling were concerned but in the area of fit and finish, something at which the Japanese were past masters, Bultaco was felt to be wanting. Mind you, it certainly added to the rugged character of the machine and was not something that was of real concern to those who wanted to own a world enduro winner.

And where the Japanese manufacturers offered a huge range of products—they were now the world's leading motorcycle nation having taken over the mantle from Germany—the Bultaco range was limited to bikes with engines in the 125cc, 175cc, 200cc, 250cc and 350cc classes. Interestingly, all engines were single cylinders with a high percentage of interchangeability between the whole range. Up until 1979 the largest Bultaco engine made was the 85 x 64mm bore and stroke 363cc unit that powered such memorable bikes as the El Montedero, Sherpa, Alpina, Astro, Frontera and Pursang. I guess one can only wonder how the Spanish managed to use model names like Pursang and Alpina without stepping on someone's toes. They must have been different times back then!

Industrial relations at the factory led to its closure by the Bulto family just before Christmas 1979. A cash injection by the Spanish Government saw it reopen with a much reduced workforce and, sadly nobody from the Bulto family. But other forces were causing the company problems and both were out of its control. The US Government instigated a raft of engine emission laws that meant the 2-stroke engine was outlawed; and the Japanese muscled in on Bultaco's off-road territory and with their vast resources soon dominated the competition.

By 1983 Bultaco was merged with Montesa and Ossa but even that venture has failed. So the famous name Bultaco had a brief but fiery commercial life over a little more than two decades. But it was fun while it lasted!

Gavin Farmer

Brooklands Books

CONTENTS

BETTER AND FASTER

Famous for their red hot racing machines, Bultaco now takes a big step in upgrading their trail and enduro machines.

The Bultaco line of purist machinery has long been highly respected for their super fast racing equipment by their competitors. This goes for both the American and European marketplaces, and pertains to just about every type of competition from motocross to trials. In the early stages, the Spanish built speedsters were as famous for their fantastic quickness as they were for their unpredictable reliability. But, over the years the Bultaco factory has improved and bettered their racing products to the point where they are now just as reliable and fast as any competition bike.

In the meantime, while all the development work was going into the racers (and perfection of the Sherpa 'T' trials bike), the road and trail machines were for the most part left dormant. For machines like the Campera, Matador and Lobito a death blow was seemingly being dealt by the Japanese, as they quickly capitalized on the dual purpose machine market with the introduction of high quality, inexpensive on-off road motorcycles. Yet, the Bultaco factory was still producing machines that were winning medals in events such as the International Six Day Trials, the world's toughest and most demanding enduro.

The development work put out by Bultaco in competing in the I.S.D.T.

The only brightly painted item is the gas tank. Saddle comfort is excellent. Small hubs and alloy rims keep the weight down.

Lighting is superb with new Fesmatronic ignition system. Speedo is easily read and resets to zero. Mounting prevents internal damage.

soon had its effect on their long in tooth dual purpose machines. A short while back we road tested the new Matador and found it to be a highly improved motorcycle, with enough improvements to make it a highly effective on-off road and enduro bike. This month we have taken the opportunity to test the Matador's little brother, the 125 Lobito.

Like the Matador, the Lobito has gone through a great deal of development to make it a much more formidable mount for the American market than its predecessor. Appearance wise the Lobito isn't any beauty queen. Other than the lighting shrouds, the only painted item is the fiberglass gas tank. The quality of the glasswork is good, although the painting leaves a lot to be desired. The fenders and side covers are aluminum and here again the finish is fairly good. In the exterior finish department the Lobito may not be a match for some of its Japanese counterparts, but again the machine is constructed for functional operation and not the showroom floor. However, this is something we would like to see improvement on (maybe next year?).

Another area that has seen some improvement is rider comfort. The saddle is a thickly padded vinyl covered rump rest that is as good as any we've tried. The saddle is long enough to allow the rider to easily shift himself fore and aft to meet various situations. There are also buddy pegs attached to the swing arm for carrying two up. (Though we would recommend the Lobito for solo riding only.) The saddle is only just long enough for another person, and riding two up is uncomfortable for both.

The riding position (solo) is fine, with all of the controls within easy reach. In saying this we must disregard the shape of the handlebars. The bars are nice and wide for leverage, but they are too far forward. Regardless of the position you put the handlebars in, they just never seem to be comfortable. This is easily remedied. Another thing we don't care for is the small diameter hand grips. They seem to be standard equipment on most European machines, but would be far more comfortable if they were larger in diameter. In the area of rider comfort, one nice thing about the Lobito is the low saddle height (from the ground). At 31 inches it's comfortable for a pair of long legs, yet just right for any of the shorter species in getting your foot on the ground without leaning the machine over at an angle.

Now and then we have the opportunity to put a machine through more than its share of grief and see just how it will fare under the most grueling of situations. Such was the case of the Lobito. On our first outing it just so happened that we were going out for a weekend of snooping in the rough and rugged Panamint Mountains in Death Valley. Here we started out in a sandy dry lake with an elevation of only a few feet and climb to over 12,000 in a distance of only twelve miles. The terrain ranges from deep sand to huge boulders and 120 degree desert heat to cool spring fed streams. The trails are all old mining roads and burro paths with the rate of ascent being super steep.

With this as a format we figured that a lot of pushing would be in store for us as we had been up here before on larger displacement machines, and we knew the severity of the climb. At the day's end we found great respect for the power and tractability of the 125 Lobito. At first we felt that the power range was much too high (RPM) and would detract from its climbing ability. Though the power range is a little higher than we would have liked, we found that the engine simply refused to stop pulling, though only slightly, from an idle on up. With the five speed gear box (with standard gearing) there was always a lower cog to grab for to get that extra bit of bite. To add the frosting on the cake, the tractability of the rear knobby was such that it never stops putting power on the ground.

The engine of the Lobito is basically a scaled down version of the larger 250cc

The exterior of the new Lobito has seen little change. Quality and finish are fair. Handlebars should be swept back further.

Weak spring strength permits the forks to continually top out and lead to premature seal failure. Spokes are hefty.

powerplants. This year's model has a higher (RPM) but wider power band than the earlier models. We initially felt this to be objectionable for an enduro type mount, but we became used to the power range and found that proper use of the gear box would let the machine pull the steepest of hills at slow speeds. Another nice feature is that the engine starts very easily hot or cold, and little effort is required to push the kick arm through its full stroke. The horsepower rating is set at 18.4 at 8,500 RPM, which we feel is an optimistic figure. But the most important thing is that the power does get on the ground and isn't lost in spinning the rear tire.

The gear box and clutch is another area that has seen a lot of improvement over the years. The gears are quite hefty for a 125, and are as large as ones seen in some machines twice its size. It's doubtful that any punishment, short of pure destructive abuse, would damage the gear box. The clutch is a multi-plate assembly that bathes in an oil supply. Here again we gave it more than its share of torment without having the slightest bit of trouble.

The gear ratios are spread just enough to permit full use of the engine's power range, yet close enough to permit grabbing a lower cog in that moment of need. The shifting was smooth and precise, and the throw of the lever is short enough ensuring a positive gear change each time. Only on rare occasions did we find ourselves in a false neutral, and this was usually our fault. The five speed gear box allows the Lobito to be plunked along at walking speed or dropped into fifth and zipped down a fire road at sixty plus. All in all,

Complete tool kit is very handy. Expansion chambe muffler is attractive and quiet. Unit does not include spark arrestor.

the versatility of the gearing and power range makes the Lobito an ideal trail, woods or enduro mount.

Probably the most notable single improvement in the new Lobito is the electrical system. For many years the poor electrics were the greatest downfall of the Bultaco dual purpose machines. Now they have made a wise decision to incorporate the use of the Fesma electronic ignition and electrical system in all their trail machines. This Fesma system is completely waterproof (no ignition points) and produces more than ample current for the complete electrical system. This new electrical system has improved the starting, overall engine performance, dependability, and most of all the lighting of the Lobito.

Just to prove our point we took the Lobito out one dark night and entered a night enduro. This is where the first rider leaves at midnight and finishes at five in the morning. The night we rode was completely absent of any moonlight at all (oddly enough the event was called the Moonshine Enduro). As the Lobito was running in with both Japanese and other European machines we were able to see just how well the lights worked. As we found out, there wasn't a machine out there with any better night visibility. In fact one nice feature of the Lobito is that the lighting is wired up directly to the battery. Therefore, when that moment of panic comes (as it did many times) and we locked up the brakes and killed the engine, we were

Firestone knobby provides maximum traction, but rubber and carcass are too soft. Rear suspension is not firm enough.

The new 125 Lobito has more steam in the mid range and has little trouble pulling the steepest of inclines without any complaints.

Buddy pegs are standard. Chain tensioner (same as Sherpa T) keeps everything in place and on the sprockets.

not left suddenly in complete darkness skidding down an unknown trail. With the lights hooked to the battery they stay on if the engine is running or not.

Another interesting facet that came to light while riding the night enduro was the handling. Riding cross-country at night (lights or not) can be learning to ride all over again. At night the terrain takes on completely different shapes and dimension, therefore the machine and rider find themselves closer to the edge more often than in daylight. We continually found ourselves falling in hidden holes, bouncing off of dark rocks and the like. As we soon found out, the suspension on the Lobito leaves something to be desired. The front forks are too spongy for rough off road use. The spring tension is too soft, and the front tire would continually come up and rub on the underside of the fender. Preloading the springs with a booster would certainly help. Also in a short period of time the fork seals began to leak excessively, and soon ruined the good damping action of the forks. At the rear we also found that the shocks are not suited for anything more than easy trail riding. Both the spring strength and damping are not up to the task of steady off road use. Improvements in these areas would surely aid the comfort and handling.

The overall handling is adequate for a trail bike, but you do not have good control at higher speeds in the rough. The wheelbase is just over fifty inches and makes the Lobito fairly docile at slow speeds. At faster speeds, above 35 mph or so, the short wheelbase, soft suspension and choice of front tire start to work against the rider. Here we feel that another two or three inches of wheelbase would help the overall handling.

At the rear is a 3.50×18 inch Spanish Firestone knobby, which is the first one like it we've seen. The type and size of the tire are good, but the construction of the rubber and carcass are soft. The softness of the rubber lead to excessive premature wear of the knobs, as they were one third gone after a mere 100 miles. Also the softness of the carcass (combined with the rigid alloy rims) lead to a couple of pinched tubes and flat tires from hitting rocks and the like. At the front is a 2.50× 21 in. block type tire that we didn't care for at all. With this size and type tire we found it very hard to control the direction of the machine on both hard and soft surfaces. On the hard stuff, like trails and roads the front end wanted to slide out in turns, and in the soft terrain the tire tended to knife in and not respond to the rider's efforts. A switch to a larger diameter knobby type would give a lot more traction and control at the front.

One newly designed item that the Lobito makes use of this year is the expansion chamber/muffler exhaust system. Exterior appearance tells you that it's an expansion chamber, but when the machine fires up the exhaust noise is completely muffled. It looks neat and clean, it's quiet and is well tuned to match the engine performance. The only thing we would like to have seen here is an approved spark arrestor system (next year?). The Lobito also comes with a complete tool kit which is something we always like to see.

Sticking with tradition, the Lobito comes standard with alloy rims to keep the strength up and weight down. The large spokes lace the rims up to the small alloy hubs. The braking fore and aft are more than adequate in the rough and sufficient for limited road work.

The Bultaco firm has been making motorcycles for the enthusiast for quite a few years now. As each year goes by they make better machines with the quality and dependability being as important as absolute performance. The new Mark IV 125 Lobito is an easy and pleasant machine to ride. It starts easy, runs quietly, is clean and smooth, and with a few changes would find its way to the top of its field. Most importantly this is probably Bultaco's best effort yet at producing a machine for the average trail rider, and not just the adamant enthusiast. — *Dave Holeman*

BULTACO 125 MARK IV LOBITO

ENGINE

Type	single cylinder, piston port, two stroke
Bore and stroke	51.5 x 60 mm
Displacement	125 cc
Compression ratio	14:1
Max. horsepower	18.4 at 8,500 rpm
Max. torque	11.15 ft. lbs. at 7,500
Ignition	Fesma electronic
Carburetion	25 mm Amal concentric
Lubrication	pre-mix gas/oil

DIMENSIONS

Length	75 in.
Seat height	31 in.
Wheelbase	50.8 in.
Ground clearance	11.4 in.
Dry weight	221 lbs.

WHEELS AND BRAKES

Front tire size	2.50 x 21 in.
Front brake type	internal expanding
Rear tire size	3.50 x 18 in.
Rear brake type	internal expanding

TRANSMISSION

Type	constant mesh, 5 speed
Clutch	wet multi-disc
Internal gear ratios	1st .287:1, 2nd .442:1, 3rd .625:1, 4th .821:1, 5th 1.0:1
Final ratio	9.12:1
Countershaft sprocket	14T
Rear wheel sprocket	54T

PERFORMANCE

Indicated highest one-way speed	67 mph
Quarter-mile acceleration:	
Top speed	67 mph

GENERAL

Air filtration	washable foam
Battery type	6V, 5 AH

CAPACITIES

Fuel tank	2 gal.
Fuel reserve	.25 gal.
Gear box	20 oz. plus 10 oz. for clutch

FRAME AND SUSPENSION

Front suspension	telescopic double damping
Rear suspension	spring over shock
Frame type	single loop tubular
Steering damper type	friction

COLORS: Red

PRICE AS TESTED: $696.00

DISTRIBUTORS

Bultaco Western
10929 Chandler Blvd.
North Hollywood,
California 91601

Bultaco American Ltd.
2765 Scott Blvd.
Santa Clara,
California 95050

Cemoto East Import Co.
PO Box 29
Schenectady,
New York 12301

CYCLE WORLD IMPRESSION

BULTACO LOBITO MARK IV 175

"A Wolf In Sheep's Clothing..."

FAST, AND BLESSED with a marvelous assemblage of components, the sum total of the new Bultaco 175-cc Lobito leaves us puzzled. It is hung high like a woods bike. It is equipped like a woods bike. But it doesn't always behave like a woods bike.

What happened is that the five-speed engine, with the high revving power characteristics of Bultaco's rapid 175 Sherpa TT scrambler, has crept into an enduro-type frame. Quoted horsepower figure for this 60.9 by 60mm two-stroke Single is 22 bhp at 8000 rpm. This is greater than the claimed output of several successful 250-cc enduro machines, which are in milder states of tune and develop peak horsepower in the 6000-to-6500-rpm range. The result with the Lobito 175 engine is that it is primarily suitable for fast riding. It is not particularly "peaky" in terms of throttle response, but the power flattens out somewhat below 4000 rpm. Thus, slogging up steep hills, or through slow, tight going requires an agile throttle hand and hair-trigger balancing.

The frame is a single-backbone, single-downtube affair, which splits into a double cradle at the engine cases. It frames the engine in a compact manner, which promotes both strength and rigidity. The tires, fitted on alloy rims, are full-size woods bike fare: a 3.50-18 Firestone Motocross at the rear, and a 2.75-21 Pirelli trials pattern tire at the front. Thus, ground clearance is a cavernous 11.5 in., excellent for ISDT-style stump jumping.

At this point, the 175 Lobito reveals another facet of its split personality. The footpeg height is a vertiginous 13.0 in.–quite appropriate to trials riding in a standing position. However, this foot height proves somewhat cramped when the rider sits down, just as it does on a trials machine.

In spite of these drawbacks, the Lobito has some excellent attributes. The front and rear suspension work beautifully, offering copious travel and good damping. The fenders are high-mounted to prevent mud blocking; these and the attractive side panels are made of aluminum. Constant chain tensioning is assured by a spring-loaded tensioner.

Fully applicable to enduro work is the large, clearly numbered Veglia-Bressel resettable speedometer/odometer. It is suspended in a rubber "cat's cradle" to isolate it from off-road shocks. Lighting equipment and horn are standard equipment, useful for short sections of pavement where the law casts a watchful eye on enduro competitors. The silencer works, after a fashion, but the noise level is still rather high, should the rider want to amble through a quiet town.

On the fireroads, the 175 will accelerate in a shockingly quick manner and run at about 65 to 70 mph, and, if it had higher gearing, about 10 mph faster. But if we are correct in assuming that its purpose is for all-terrain riding, a meeker engine with mild port timing, and footpegs mounted about 2 in. lower, would help this latest Lobito do its job better.

CYCLE GUIDE

TEST REPORT

BULTACO 360 EL MONTADERO

BULTACO'S BIGGEST (AND BEST) ENDURO BIKE

After having more than their share of miseries with their first 360cc Montaderos, the factory re-grouped for a think session. Bultaco's new big bore enduro mount shows more than just casual improvement.

For nearly a decade, Bultaco has been one of the pioneering manufacturers of first-line competition machines.

Their Sherpa S and Pursang machines have won more than their share of races in this country. The reputation of the Spanish factory has rested on the record of the internationally successful Matador and Sherpa T machines, all first-class, competitive bikes in every detail.

But somehow, their big bore (over 250cc) machines, it seems, like the 360cc El Bandito and El Montadero, never really cut the mustard.

Doubtlessly, a great deal of soul searching preceded Bultaco's decision to improve it. After testing the new El Montadero, this magazine is glad to report a much needed improvement.

As serious, big-bore enduro mounts go, the Montadero is one of the most durable and professionally constructed machines available. Consistent with its 360cc displacement, the Montadero is a hefty-looking machine.

Bultaco uses large, fiberglass gas tanks and side covers on this and many other models. The finished quality is good. The painting, though, is not. The front and rear fenders are heavy-guage, chromed steel that

Not what one would call pretty, the new El Montadero is a functional and rugged enduro mount. The overall finish is fair, but construction is strong.
The double damping front forks give an excellent ride and perform very well. The 21-inch knobby is well matched to the machine. Lighting is very good.

should last. Not intended to be a lightweight, the chassis is a large, heavy assembly. No amount of punishment will phase the chassis. All welding was planned with an eye towards good appearance and strength. Overall, it's a handsome, but not beautiful, package.

Bultaco designed the Montadero as a serious enduro competitor, a bike that would successfully beat any challenger or condition. There's no question the Montadero has all the equipment to do that. One prime advantage is its big bore-muscle. When properly used, the Montadero's power can pull a rider over obstacles that would stop smaller bikes.

The engine is probably the smoothest and most predictable big-bore, enduro power plants tested by *Cycle Guide.* The power comes on at very low speeds and pulls steadfastly through the higher gears. The power band is very low, resembling a trials engine and runs best from about 2,500 to 5,000 rpm. This is perfect for enduro riding.

One external part worth noting is the torque-designed feature. The Montadero has the longest headpipe we have ever seen on a two-stroke bike. Few big-bore head pipes exceed 20 inches in length. The Montadero's is a fantastic 34 inches.

Starting many big-bore, two-stroke singles usually requires a good deal of effort, but not so with the Montadero. Since the kick arm is on the right side of the machine, a rider just straddles the bike and pushes the arm through. Even a smaller rider can get the fires burning quickly. When cold, this reluctant Latin engine needs generous tickling of the 32mm Spanish Amal concentric. More often than not, the engine has to be started several times successively to keep it running because there is no choke. When warm, the carburetor needs a slight tickling and the bike will start on the first try.

A rider not familiar with the Montadero will think there is something wrong when he starts the bike. There are no clattering or whinning noises and the exhaust popping is barely audible. If other motorcycles are running nearby it's difficult to believe the Montadero engine is even running. The lack of noise makes it a very enjoyable bike to ride. This is the quietest big-bore, two-stroke single we have ever tested. We wish all dual-purpose machines could equal the Montadero's low-noise level rating.

We were favorably impressed from the time we started the bike, then quietly dropped the gear lever in first and let the engine pull us away with an amazing amount of low-speed power. With all its power on tap, very little throttle is required to get the machine under way. We went through the gears smoothly and quietly, although the throw of the shift lever was excessive. Moving the lever up or down is smooth and precise. The clutch wasn't as smooth as we would have liked, but it never really caused any problems.

Getting from place to place regardless of the terrain is what the Montadero is all about. It does this with little strain on the engine and rider. The amount of power pumped by this engine makes it unnecessary to ever rev the engine above five or six grand. The rider can feel its rev limit through the vibration in the handlebars and foot pegs. This isn't to say that the

The robust oversquare engine has an amazing amount of muscle. Carburetion—an Amal concentric—performs well. Dual petcocks are located on the same side of the tank.

One thing we didn't care for at all was the clip-on type handlebars. Fortunately, conventional bars can be mounted. The rear-wheel-drive speedo is easily read and resets to 0.

In an effort to acquire as much low-speed torque as possible, the El Montadero uses this super long head pipe. It looks like it belongs on a four-stroke, but it works.

The Montadero is all new, fore to aft. New chassis design improves the handling. The large gas tank and side panels are 'glass, while the fenders are durable chrome steel.

engine shakes. It doesn't. The rider just becomes attuned to little things like the velvety engine vibrations when there is no loud exhaust noise.

A rider knows he's on a hefty machine the first time he throws his leg over the saddle and rests on the Montadero's thick, black seat next to the comfortably wide gas tank. One strange thing is the foot peg location. Unlike most off-road machines that tend to force the rider's knees high, the Montadero has a low foot peg position. After several hours, this becomes an unexpected blessing. We found that the foot pegs are about two comfortable inches lower than other off-road machines. The only problem we found with the seat is that it tends to bunch up towards the rear and force the rider to slide forward into the gas tank.

The stock clip-on style handlebars that attach to the top of the front forks are uncomfortable, but furtunately, the top triple clamp permits bolting on conventional one-piece handlebars.

The new Montadero's chassis and geometry has been changed from last year's model. This was needed. Bultaco's chassis alterations improved its handling considerably.

We found that charging hard and fast over rough terrain is much easier and more predictable with the new model. But because it is an enduro machine, it still does not perform well at high speeds over desert terrain. Keep in mind the Montadero wasn't designed for high-speed desert riding, but it excels in the slower, tighter, rougher going. Under ideal conditions, the Montadero becomes an extension of the rider, performing to the limit of his skills and not being the limit of them.

The Montadero is no featherweight at 285 pounds, but virtually all this weight is in its indestructable superstructure. Despite this weight, the machine is very docile. The Montadero seems to be a bit nose-heavy, though. Getting its 21-inch front wheel off the ground requires effort and planning. We discovered the Montadero is much better riding over objects than wheeling over them. The heavy front end makes the bike stable going down hills and riding on fire roads at high speeds.

The Montadero's traction is good, mostly because of the engine's torque. The suspension fore and aft is cushiony and both the front forks and rear shocks are soft, giving the rider a very shock-resistant ride. This is another reason the Montadero handles nicely at slow speeds. The spring rate and damping of both suspension units are well-suited for enduro conditions.

BULTACO 360 EL MONTADERO

ENGINE

Type	single cylinder piston port two stroke
Bore and stroke	85×64 mm
Displacement	362.8cc
Compression Ratio	10:1
Rated Max. Horsepower	32.6 @ 6,000 rpm
Rated Max. Torque	36 ft. lbs. @ 5,000 rpm
Ignition	electronic CDI w/battery, dual plug
Carburetion	32 mm AMAL (Spanish) concentric
Lubrication	premix gas/oil

DIMENSIONS

Length	84.6 in.
Seat height	31.0 in.
Wheelbase	55.9 in.
Ground Clearance	10.4 in.
Dry weight	279 lbs.

WHEELS AND BRAKES

Front tire size	3.00×21 in.
Front brake type	internal expanding
Rear tire size	4.00×18 in.
Rear brake type	internal expanding

TRANSMISSION

Type	constant mesh 4 speed
Clutch	wet multi plate
Internal gear ratios	1st .405:1, 2nd .645:1, 3rd .925:1, 4th 1.26:1
Final ratio	8.148:1
Countershaft sprocket	15T
Rear wheel sprocket	42T

PERFORMANCE

Indicated highest one-way speed	94 mph

GENERAL

Air Filtration	dry paper
Battery type	6V-5AH

CAPACITIES

Fuel tank	3.4 gal.
Fuel reserve	.5 gal.
Gear box	600cc

FRAME AND SUSPENSION

Front suspension	telescopic double damping
Rear suspension	adjustable spring over shock
Frame type	tubular single cradle

COLORS

Red

PRICE AS TESTED

$1,096.00 F.O.B. Santa Clara

DISTRIBUTOR

Bultaco American, Ltd.
2765 Scott Blvd.
Santa Clara, Calif. 95050

DEVELOPED AND PERFECTED by Sammy Miller, unquestionably the best trials rider in the world, the 1971½ Bultaco Sherpa T abounds in new and subtle changes which make it quite possibly the best trials machine in the world.

The most obvious change from the 1971 model is the Kit Campeon fiberglass work, which consists of the narrowest seat/gas tank combination to be found on a 250-cc trials machine. A mere 3 in. wide at its barest width, and not more than 5 in. across all the way up to the front of the tank, the Kit Campeon offers ease of control and maneuverability while clambering around on the machine (maintaining one's balance) that has never before been realized. Even though the exhaust header pipe protrudes outside the bottom of the tank/seat on the right hand side, there is no contact between it and the rider's leg.

The gas tank capacity is 1.2 U.S. gal., which is more than adequate for a loop in any observed trials event, and a breather tube from the top of the filler cap is routed down through the front fork stem, which prevents gas from spewing out on the ground if the machine is allowed to fall. The extremely narrow seat has an average width of 6.5 in., providing only a modicum of support for the buttocks while traveling between sections. But sitting down isn't the proper way to ride the Sherpa T. It's a stand-up bike from the word go.

Although primarily designed as a trials bike, only about 30 percent of the Sherpa T models sold are used exclusively for trials competition. Many owners use them for cowtrailing, and a few use theirs in enduros. For this customer, each Sherpa T comes with a headlight and a taillight which can be installed in about 20 min. to make the bike "enduro legal," but not legal for highway use.

Front forks are Betor units, built to Bultaco specifications. They offer nearly 7 in. of travel, and have a soft spring rate. Damping characteristics are excellent and can be changed by substituting oil of heavier or lighter viscosity. Robust as they are, weight is kept to a minimum by using thin-wall tubing for the stanchions and aluminum for the top and bottom crowns, as well as the lower legs.

Rear suspension chores are admirably handled by five-way adjustable Betor shock absorbers which offer nearly 4 in. of travel. It's practically impossible to fault the Sherpa T's suspension operation, but leaking front fork seals pumped oil so badly that they had to be replaced almost immediately. Since then, however, the fork tubes have remained oil free. Rubber slider wipers prevent the ingress of dirt into the seals.

Somewhat puzzling is the front fender's proximity to the front tire. In areas where clay is not a main constituent of the soil, the fender's location is not a problem in wet weather. In certain areas of the U.S., however, mud tends to stick to the front tire and builds up between it and the fender, locking the wheel. Of course, with the fender mounted close to the front tire, it is easier to see exactly where the wheel is going, but raising the fender another inch would help matters in wet going.

Following traditional Bultaco practice, the frame is a single toptube, single downtube design with the downtube splitting into two smaller tubes just in front of the engine. These two tubes continue rearward and terminate at the rear suspension top mounting points. A small loop extends rearward from there to provide a method for attaching the rear fender.

CYCLE WORLD ROAD TEST

BULTACO SHERPA T

The Kit That Is No Longer A Kit Makes The "T" Even Better.

The engine is secured to the frame in three places and is protected by a perforated skid plate welded between the bottom cradle. Above and behind the crankcase, and just above the swinging arm pivot gussets, a smaller tube stretches from the rear downtube back to the rear cradle members, forming a strengthening triangulation that provides a mounting point for the front of the rear fender. The swinging arm pivot gussets do double duty by serving as the footrest mounting points.

Whatever other virtues the complete Sherpa T possesses, the engine/driveline package has got to be the smoothest available for trials riding.

Low compression ratio, mild port timing, a small 27-mm Amal concentric carburetor, and inordinately heavy flywheels make the machine a low-speed "lugger" without equal. To provide the best possible spark at ultra-low rpm, a flywheel magneto is employed instead of the now popular Femsatronic ignition. A spring-loaded primary chain tensioner and a similar one for the rear chain keep all slack out of the driveline, making power delivery instantaneous and smooth. This instant response can make the difference between cleaning a trials section and losing a mark for footing.

The primary drive is interesting in that a huge single flywheel is mounted on the outer end of the drive side of the crankshaft, in addition to the flywheel magneto on the opposite side and of course the ordinary crankshaft flywheels. The auxiliary flywheel is V-grooved near the center with a sprocket machined at the bottom of the groove for the single-row chain to the clutch. Ideally, the clutch should only be needed infrequently in a trials event, but the clutch is the same as is used on the powerful Pursang scrambler, and will tolerate almost any amount of abuse.

In addition to the power-producing portion of the engine unit, the gearbox is also highly specialized. First, second and third gears are spaced very closely, for use in the most difficult portions of a trials course. Fourth is a good intermediate gear and fifth is only needed when traveling between sections at speed. This may not sound important, but trials in Europe are run on an elapsed time basis, which not only necessitates making good time *through* a section, but also *getting to* the next section as fast as is practicable. Shifting action is exemplary in spite of the rather short gear lever, and a very sophisticated clutch rod ball and roller bearing set-up adds to >

the smoothness and reliability of the unit.

The tiny front brake and normal rear brake provide more than enough stopping power for trials riding, and are less apt to get wet because of longer sealing lips and more careful machining on the brake backing plates. Thick, strong spokes connect the hubs with Akront alloy rims which are somewhat lighter, and allegedly stronger, than steel rims. In typical trials fashion, the rear tire is a 4.00-18 and the front is a 3.00-21. Not so typically trials is the tread pattern: knobby! Each factory distributor has the option of tire tread patterns, and Bultaco services orders their Sherpa T models with knobbies.

This may seem curious, as Trials Universals are considered the usual tread for observed trials. In Europe and Great Britain, the universal tread is required by FIM rules. However, the situation is different in the U.S.; most trials clubs here allow use of the knobby pattern, except in certain areas (notably Southern California, where the universal is required). And not forgetting how many trials bike owners do not ride trials, it is small wonder that the distributor opts for knobby tires.

Location of the motorcycle's controls is conventional for a trials machine with flat, wide handlebars, a low seating position and high footpegs. The rear brake pedal is interesting in that it pivots from inside the frame tubes, which helps prevent breakage in rocky terrain. The actual point of contact between it and the rider's left boot is circular in shape, hollow to prevent the build-up of mud, and features sharp, saw-toothing on top for better grip. The footpegs are spring-loaded and similarly saw-toothed.

The gear lever on the right side is sufficiently far away from the footpeg to prevent snicking the transmission into a neutral inadvertantly while negotiating a tricky section. Clutch and brake levers are adjustable steel items, which seems strange in view of the apparent effort to save a few ounces by the use of an aluminum alloy transmission drain plug. Why not use aluminum levers instead?

The 1969 Sherpa T test weighed in at 219 lb. wet, whereas the 1971½ Kit Campeon tips the scales at 212 lb., the lightest 250 trials bike available. Finish and quality of construction have come up considerably in the past few years. Although a couple of the frame tubes are a little rough, the assembly's welding is very good indeed and little touches here and there are pleasing. Some of the little touches are annoying too; for example, the muffler's "stinger" should have been made round instead of tapered, and angled slightly outward to facilitate the installation of a spark arrestor. Now, the exhaust blows onto the speedometer cable, cooking the insulation off, and sprays an oily mist onto the rear hub. All the fender mount bolts and those three which attach the seat/tank unit should have elastic stop-nuts as standard equipment. It is also necessary to re-torque the top fork nuts to 150 lb.-ft. to assure proper seating of the fork tubes, and there are other little things that need to be done to make the machine 100-percent reliable. Of course, *any* bike designed for dirt riding should be thoroughly inspected and adjusted before it is ridden seriously.

In spite of some small criticisms, the Kit Campeon Sherpa T is the finest 250 trials machine presently available.

BULTACO SHERPA T

SPECIFICATIONS	
List price	$895
Suspension, front	telescopic fork
Suspension, rear	swinging arm
Tire, front	3.00-21
Tire, rear	4.00-18
Engine, type	two-stroke Single
Bore x stroke, in., mm	2.83 x 2.36, 72 x 60
Piston displacement, cu. in., cc	14.9, 244
Compression ratio	9.0:1
Claimed bhp @ rpm	19.8 @ 5500
Claimed torque @ rpm lb.-ft.	N.A.
Piston speed (@ rpm) ft./min.	2172 @ 5500
Carburetion	27-mm Amal concentric
Ignition	flywheel magneto
Oil system	oil mist
Oil capacity, pt.	oil in fuel
Fuel capacity, U.S. gal.	1.2
Recommended fuel	premium
Starting system	kick, folding crank
Air filtration	washable foam

POWER TRANSMISSION	
Clutch	multi-disc, wet
Primary drive	single-row chain
Final drive	single-row chain
Gear ratios, overall:1	
5th	8.79
4th	12.12
3rd	19.86
2nd	25.68
1st	33.49

DIMENSIONS	
Wheelbase, in.	52.3
Seat height, in.	30.0
Seat width, in.	6.5
Handlebar width, in.	33.3
Footpeg height, in.	13.8
Ground clearance, in.	11.0
Curb weight (w/half-tank fuel), lb.	212
Weight bias, front/rear, percent	45.5/54.5

Riding the BULTACO MATADOR 250 in Spain

As I was due to spend a few days in Barcelona, reporting on the motor car grand prix for *Motor Sport*, the thought occurred to me that the last thing I wanted in a hot steaming city like the Catalan capital was my E-Type Jaguar. The Barcelona race is held in Montjuich Park at one end of the main avenues, and I was staying about mid-way along its length, so like everyone else I visualized using taxis for getting about and thus avoid all parking non-senses and the need to learn the best routes about town. Then I remembered that the Bultaco motorcycle is built in Barcelona, and Mr. Bulto and his family are old friends of mine from more than 20 years ago, when I raced a 125 c.c. Montesa for him on the Montjuich Park circuit, so I let him know when I was arriving and he promised to lend me a motorcycle for getting about the city and to the circuit. More and more I find myself sitting in traffic jams in the Jaguar wondering what I am doing there when I could be easing in and out of the traffic on a motorcycle. Of course, if you can't ride a motorcycle you are doomed to remain in the traffic.

A visit to the Bultaco factory is an experience in itself, for it is hidden away in the suburbs of Barcelona in what was once a farm, and is now surrounded by factories and blocks of flats, and the approach to the factory is down a rough track alongside some railway lines. There is no large name plate to indicate you are at the home of the Bultaco motor-cycle, but you are more than likely to meet a Bultaco Pursang moto-cross bike in full flight on trial along the rough track, or to see one hurtling along the railway embankment. My taxi driver knew exactly where San Adrian del Besos was, and thought he knew where Moto-cicletas Bultaco could be found. He was not far out, just the wrong side of the railway lines, which was no problem for Spanish railways are open to the winds, so all I had to do was walk across the lines and in at the front door. I had arrived a day early (E-Type motoring?) so Mr. Bulto explained that I could borrow his "taxi" for the first day and change it for a new bike the next morning. The "taxi" is a little 200 c.c. black and grey road bike that he has been using for the past eight years for commuting to the factory when traffic is bad or for making a round of business calls in the city, the time saved by having no parking problems being incalculable, apart from the complete lack of traffic frustration. Before leaving on the "taxi" I watched Malcolm Davis have a blast up the railway lines on a works 250 c.c. moto-cross bike, for the competition department was all agog getting ready for the Spanish 250 c.c. moto-cross world championship meeting, unfortunately to be held at the same time as the car grand prix.

Next day I swopped the "taxi" for a 250 c.c. Matador Mk. IV and Mr. Bulto said he particularly wanted me to sample

"That it had been developed by the International Six Days Trial was self-evident in its appearance . . ."

the latest Matador as he and his design staff were very proud of it, feeling that it represented a very good all-round motorcycle that would do anything and go anywhere, with reliability and practicality bred in it from the factory's efforts in the International Six Days' Trial.

During the next few days I was able to put in 250 miles of the most varied uses I could find and on all of them it produced very high standards, more than coming up to the factory's claims and proving to be a most enjoyable machine. The Matador is a cross between a trials iron, a scrambler and a road-racer and is exactly the sort of bike that *Motorcycle Sport* and many of its readers have been deploring the demise of in recent months. Before the days of super-specialization in the various branches of motorcycle competition the clubman could use his bike for rallies, trials, scrambles, grass-track or hill-climbs, even short circuit racing, merely changing sprockets, tyres and handlebars. Nowadays even to get an entry in those activities you need a specialized bike, so the all-round rider would need at least five machines. Not everyone wants to take part in motorcycle competitions, but I am sure that everyone would like to have a dabble at the various things you can do on a motorcycle, to try and do a Sammy Miller on the mud and rocks, or a Joel Robert over the rough stuff, or to have a touch of the Chris Pusey on someone's field, even to ape Paul Smart round some of the corners on the open road; I know I do and I found the Bultaco Matador would do all these things in a most satisfactory manner.

On first sight the Pirelli knobbly tyres looked as if they were going to be a handicap to "short-circuit racing" through the streets of Barcelona, but you can ignore the knobs and dash in and out of corners with no worries at all. I was genuinely surprised for I thought there would be a "washboard" effect when you cranked over on to the knobbly edges. The Matador has a five-speed gearbox controlled by a lever on the right and the gear ratios are such that you never give them a thought. There is always a gear for the moment, you just keep hooking up or down and keep the engine on the boil. It's "down for one" and up for all the rest. Using the Matador in the city solved all the problems of parking, finding the way, nosing about down little alleyways where a car would be an embarrassment, and there never seemed to be any traffic jams, for me at any rate, though there were times when I felt the whole of the city traffic had given up and parked permanently.

The grand prix practice being at the civilized time of 4 p.m. I took off early one morning to sample the Matador out where it really belonged and where it had been bred, off the road and up in the mountains. My first objective was the mountain road used for the Montseny hill-climb, some 30 miles north of Barcelona, and I used the new Spanish Motorway to get to the foothills, the Matador humming along between 50-60 m.p.h. very contentedly, and being equally happy between 60 and 70 m.p.h., its top-whack being just over 75 m.p.h.

The writer says his low "centre of gravity" is a handicap when it comes to controlling a bike in the rough

It "scurfed" its way up the 16.3-kilometre mountain hill-climb with great joy and then really came into its own when I went on up the mountain, first on cart tracks and then on goat paths that finally petered out in the wilds. Poised on the footrests, you could pound along rough tracks in scrambles style, the bike feeling nicely balanced when you took off over humps, and it was just as much at home on the "nadgery" stuff and loose rocks. With the fat knobbly tyre on the back, a low first gear and a wide range of power in the two-stroke single-cylinder Bultaco engine you felt you could go anywhere, and the all-up weight is such that even my 5ft 2½in stature could lift it out of difficult predicaments, like scoring a five on a tricky section.

I stopped at the top of the Montseny and took stock of this very compact and healthy-looking Spanish motorcycle and found much that was pleasing. That it had been developed by the International Six Days' Trial was self-evident by its appearance and the way it had romped over all manner of going to get to my stopping point. However, more impressive were the details which made you realize that the Matador was made by practical motorcyclists and not board-room people and salesmen. The handlebars are wide and comfortable, suited to trials or scrambles, and on the plastic, water-lubricated twist-grip is a ring of sponge rubber to protect the thumb. Haven't we all at some time or another cut a circle of sponge to slip over the twist-grip of our trials bike to stop chafing the thumb? Bultaco fit one as standard. The speedometer which is mounted on the right of the forks sits in a rubber suspension unit and on the left side of the bars is a multiple switch (made in Barcelona) that operates horn, lights (pilot and head), dipswitch and cut-out button, the ignition being turned on or off completely by a key under the seat. Putting all the electrical controls on one switch seems so obvious and it means that the headlamp can be tucked away between the fork stanchions without the necessity of leaving space for getting at switch gear built into the lamp.

The speedo is driven from the front hub and the cable passes through a wire guide, together with the front brake cable, to keep them out of the way of the wheel and forks when on full lock. This wire guide is standard and the sort of thing any enthusiast would add to his bike to finish the job off properly. The footrests are hinged and spring loaded, so that they fold up if you lay the plot down inadvertantly, and are rugged steel structures with toothed edges and not adorned by useless footrest rubbers like that strange moto-cross machine that is sold in Birmingham. The Matador footrests are absolutely right for the job, as is the very full steel sump shield that is fitted. A truly outstanding feature of this standard production machine is the totally-enclosed rear chain, an alloy box enclosing the rear sprocket with rubber-jointed tubes for the top and bottom rims of the chain to operate in. After initial bedding-in of a new chain the free-play is adjusted and you can then virtually forget about it. As proof Mr. Bulto and his nephew

"I took off early one morning to sample the Matador where it really belonged . . . up in the mountains . . ."

proudly explained how the works Matadors will do the whole of an ISDT without a single chain adjustment. Playing about in the dusty mountains and on some muddy going that I found, I must say it was comforting to know that the rear chain was clean and dry. Alloy rims are standard wear as are the Pirelli "comp" tyres and the whole bike is neat and compact, finished in the normal Bultaco colours of red and silver.

Before handing the Matador back to the factory I used it for some laps of the Montjuich circuit, and a dash along the winding coast road to Sitges, to the south of Barcelona.

I thought as a road bike for pleasant swervery it was more than successful, but it was not an out-and-out road burner, nor an ultra-long-distance touring bike, the high-pitched ring of the two-stroke engine not being very relaxing. As an all-round everyday usable machine with no vices that I could find, it was admirable and a true "Clubman's" machine with performance within the limits of 250 c.c. as one would expect. Its real charm lay in its ability to do anything and go anywhere, embodying all the character and spirit that makes a motorcycle something apart from other forms of transport. On any motorcycle you instinctively get a feeling of freedom, of not having to conform to a tiresome pattern simply because space is limited. A motorcycle takes up so little space on the road that congestion is unknown and this freedom extends to the cart tracks and goat tracks and the open spaces just as much as it does exploring the quays of a large port or the narrow passages of an old city. A Honda 4 or a Norton Commando are fine and impressive for the job they are built to do, but a bike that is born of ISDT experience is inevitably a better all-rounder and if you like motorcycling you must enjoy it whether you are doing 80-100 m.p.h. down the Motorway or picking your way over the rocks and streams of mountain country.

The fact that the Bultaco Matador is such a practical, enthusiast's machine is due to it being conceived by a motorcycling family, the head of which began motorcycling in the golden years of AJS, and one of his sons is at present the Spanish Trials Champion. That the family and firm have faith in the Mk. IV Matador is shown by the recent announcement that a production model is being entered for the 250 c.c. Production TT, ridden by someone like Tommy Robb, and then the same machine will be ridden by Sammy Miller in the ISDT, both events under strict observation in an attempt to gain the Maudes Trophy which is awarded for outstanding achievements by motorcycle manufacturers. If the Mk. IV Matador fails to win the Trophy it will not be due to its unsuitability for the job that is in hand.

At present the total output of the Bultaco factory is in the order of 15,000 bikes per year, of which some 4,000 a year are Matadors, the Californian market rating them high on the "trailbike" list. In Britain the Sherpa trials bike has a very big following and in Spain the little 150 c.c. touring machine is a good seller for beginners and commuters, but the Matador is the Bultaco for the enthusiastic clubman.

When I handed the bike back, with apologies for its dusty and muddy state, Mr. Bulto smiled and said it was of no importance "for," he said, "I can see that you enjoyed my Matador, and we at Bultaco think it is a very nice machine." They are absolutely right, in my opinion.

D. S. J.

NB.—The Mk. IV Matador is imported by Comerfords of Thames Ditton and sells for £365 all in.

bultaco matador
mark 4 sd
ALL THE NECESSARY INGREDIENTS

With gas in its tank the bike weighs in at a feathery 245 pounds.

Typically Bultaco, forks provide excellent dampening and wheel control, are not leak-free however.

Function, pure and simple function. Bultaco's character is predicated upon this. It must work well first, then it is time to worry about all the little stuff. All the good looks in the world won't get a stone off the starting line. The men at Bultaco know this and their machines reflect this attitude.

Take, for example, the new Matador. It's officially called the Matador Mark 4 SD. The SD stands for Six Days, y'know, like in the International Trial. This motorcycle is designed specifically for that raunchiest, kind of cross-country competition, the enduro.

The enduro, where grit and grime get in places you didn't even know you had. The enduro, where Mother Nature brings out her whip, spurs and beanie with a button on top announcing to all "Watch it, Big Boy, I'm a sadist!" Ah yes, it's such tortures as this that make a guy really appreciate Home with its ever-present mothers-in-law, foul-mouthed neighbor kids, house cats with poor bowel habits and gardeners who don't but bill you anyway. Hills, holes, mud, slime, sticks and stones; everything and anything to break your bones – such are the things enduros are made of.

But, strangely, some guys thrive on this sort of thing. Lord knows it's challenge enough for the greediest masochist. Clearly, motorcycles used in this type of competition must be equally tough. A bike is required that will take all the abuse you can give and yet neither malfunction or complain. A bike like the new Bultaco Matador.

The pedigree of this machine is quite impressive. It is, you see, almost an exact duplicate of the works Bultacos that won four gold, five silver, and four bronze medals in the 1970 International Six Days Trial. Mind you, it is not a "replica" in which a few features are shared with the original racer. This is *the* bike. The only area that differs from the factory machine is the inclusion of the A.M.A.-required enduro goodies; map holder, clock mount and speedo. In the past year, though, certain detail improvements have been made to this model. The air filter has been greatly improved, the rear brake is now rod-actuated (before, a cable did the job), and the cylinder is now six-ported for increased flexibility. The thing costs $995 on the west coast; a substantial tariff. But the motorcycle is, to our minds, well worth it. It's not every day that you can buy what is essentially a factory-prepared machine already race-proven.

Okay, so what sets this bike apart from its competition, other "enduro" motorcycles? Well, there are several

matador

things. Importantly, the entire rear chain is enclosed. The value of this feature is so obvious we find it hard to understand why more manufacturers do not incorporate them on their machines. Still, chain cases are not all that common even though they will protect the chain to such a degree that chain and sprocket life is at least doubled; often tripled and quadrupled. The Bultaco Matador line has had this item ever since the five-speed transmission was adopted. The case has been modified for the latest models with several extra pounds being eliminated.

It is particularly convenient that a tank-mounted tool kit and handlebar-mounted map holder are included as standard equipment. All too often, if a guy wants to go enduro racing he must purchase these goodies as accessories, and such pieces are seldom cheap. Additionally, the speedometer is rubber mounted, thusly vibration-proof and a mount for an enduro clock is provided.

If the number plates appear a bit odd to you, rest assured: they are. The plates are made of rubber, and for a very good reason, too. Bultaco Major Domo, Tom Patton, phases it as well as anyone else: "Because a lot of guys will eat their number plate when they high-side. Also, the things are immune to vibration and won't chip or crack in event of a prang.

In order to prevent twigs and grass from gumming up the works between the foot levers and frame rails, a wire runs between the forward end of each lever and is fastened to the frame directly ahead of the engine. Again, this is standard procedure in enduro racing but not too many enduro bikes come with it out of the box.

The best way to describe the Matador's handling is to say that a guy has got to be a real klutz to get it seriously out of shape. In a word, it is *forgiving.* While the Matador weighs in at 245 pounds wet, its handling is very similar to some of the finer handling eastern European motocrossers weighing 25 pounds less. It is eminently easy to steer by throttle, wherein you pick your line, lean the bike over and gas it accordingly. While bustling through a bend the bike exhibits the kind of confidence where you don't worry about small, mid-course corrections in steering and throttle settings.

The test, however, was not without its problems. Our first hassle is, we suspect, traceable to a faulty float setting. Throttle control at low revs was erratic, often refusing to idle

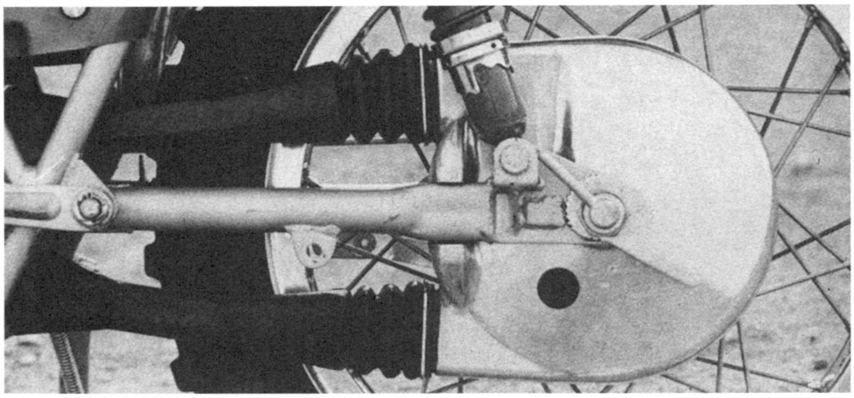

We really dig the enclosed chain. An inspection hole is provided allowing quick and easy maintenance. You can expect chain life to be doubled over that of an exposed unit.

Purpose of the wire betwixt pedal and frame is to prevent twigs and foo from being snagged therein, gumming up the works.

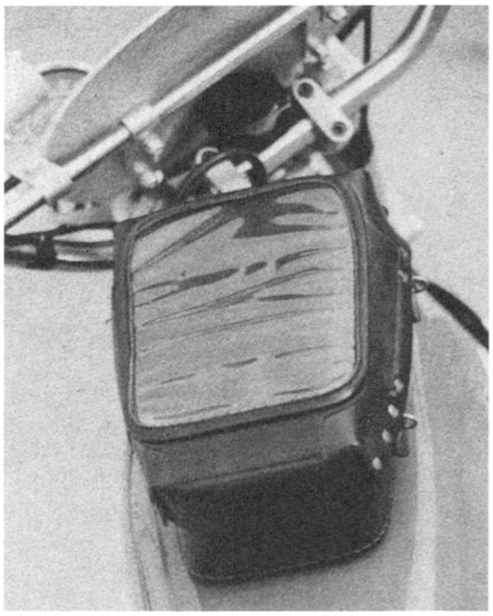

The top of the tool kit is to be used for holding check slips and notes for enduros. Transparent cover keeps things dry.

In order to protect it from the elements, carburetor is shrouded by vinyl cover.

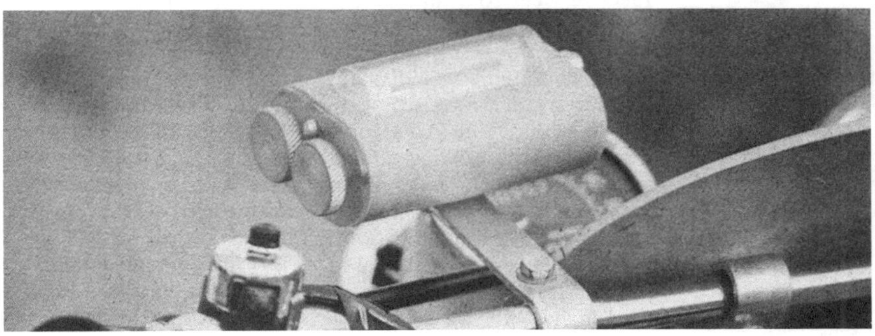

This nifty is a map holder. It is extremely handy although mounting provision is not too sturdy.

down, staying around 3500 rpm. A bummer. The throttle cable was slackened as much as possible with no effect at all. The metering needle was then lowered one notch which helped somewhat, indicating an over-rich condition. At this time also the carburetor slide was closely inspected for galling or other evidence of hanging up in the throttle bore. Nothing suspicious here. Additionally, as the engine rambled at its rapid idle the tickler button was lightly whunked a couple of times in the hope that the float was merely stuck. That didn't work. Oh well . . .

One second bugaboo was the ignition kill switch. It shorted out after two minutes riding in the rain. Of course, this won't do. The slightest vulnerability here is intolerable on an enduro machine, particularly one costing $995. Perhaps there are aftermarket kill switches available sufficiently insulated to plus this sizeable hole in Bultaco's championship bota bag.

Thirdly, as the test progressed the small map holder on the handlebar suddenly fell off. It seems that all that held it to its bracket was a small screw running into the holder's plastic body. With time and vibration a break occurred, the goodie fell away to be lost forever in the aromatic mesquite of the Santa Monica Mountains – may it rest in peace.

So there you have it, Bultaco's latest offering to the enduro enthusiast. Sure, we had minor hassles with it during our test, but we also had a lot of fun with the bike in spite of them. Too, more than a few minds were blown and eyebrows raised when the bike was wheeled off the truck at our favorite riding areas, not to mention when we blew past other riders in the stickiest, most miserable riding sections. The choice is yours. If sorting out little bugs bothers you, don't get it. But then if you're the kind of biker with no patience for such things, you're not the enduro type anyway.

BULTACO MATADOR MARK 4 SD

PRICE

$995

ENGINE

Engine type	two-stroke, single cylinder
Horsepower @ rpm	28 @ 7000
Torque	21.8 ft./lb. @ 6000
Bore and stroke	.72mm x 60mm
Displacement	244.3cc
Compression ratio	9.0 to 1
Carburetion	32mm Spanish Amal
Ignition	Femsatronic pointless

TRANSMISSION

Speeds	five
Clutch	multi-disc, wet

CHASSIS

Length overall	81 in.
Wheelbase	54-3/4 in.
Ground clearance	10 in.
Weight	245 lb.
Frame type	single down-tube
Tire size, front	3.00-21
rear	4.00-18
Brakes, front	single leading-shoe
rear	single leading-shoe

THE FIRST LOOK INSIDE THE ALL-NEW BULTACO RACING ENGINE

INSIDE THE 350 BULTACO PURSANG

John Larsen

What's in the 350cc Pursang? This motorcycle turned up as quite a surprise in the last Trans-AMA moto-cross series as everyone had assumed Bultaco would run their 360 Bandido in the open class. Much to everyone's amazement Tom Rapp came out to battle the open class with this new secret weapon from Spain.

The assignment of the 326 Bultaco to Tom Rapp, and also Bultaco rider Gary Bailey, was to not so much try for the championship honors, but rather see if they could break the bike because the two cycles were prototypes for a planned production model. The riders gave the machines everything they had, and the machines were shipped back to the factory, but before they left we persuaded Tom Patton of Bultaco Services to unbutton the engine from Rapp's bike and let us peer over his shoulder while the engine was dismantled for a quick look-see.

We whipped into Patton's combination office-garage-test station and found Tom had the engine ready for inspection on the engine stand. Not only were we interested in the innermost secrets of the engine but we were also curious to see what shape it was in after a great deal of all out moto-cross time had been put on the engine.

Tom removed the head and a big change could be noted as soon as it was removed. For openers, the intake and exhaust ports were both unbridged, a departure from past Bultaco tuning. Tom explained that the bridge was found to create unwanted turbulance in the larger displacement engine. The transfer porting was of the two on each side variety, giving the upper end of the engine a total of five ports, including the exhaust. The cylinder itself is of the square design, much like the Mk-V Pursang 250. Carburetion was handled by a 34 Amal Concentric. It's the new design with the relocated air idle circuit to give cleaner operation off the bottom end. The production model has a 36mm carb. The bore and stroke of the engine is 83.5mm by 60mm, giving an actual displacement of 326cc.

We had heard that the 350 used the same lower end (crank) as the 250cc machines, but a quick look at the wrist pin end of the rod proved immediately that this was not the case because the 350 wrist pin was 3mm larger in diameter than that of the 250.

To get to the bottom of things, Tom split the cases so we could check the bearing size at the bottom of the connecting rod. The cases come apart the same as with existing Bultaco five-speed engines, but are not the same. The area on each side of the cases are beefed up a good 2mm deeper than the 250, and of course the spigot area where the large barrel fits the cases is machined out over the 250cc size.

When the cases were parted, we found that in no way is the same crank used, as the flywheels of the 350 crank are 2.5mm larger in diameter than the 250. To try and put a 350 crank in the 250cc cases just wouldn't work.

The main bearings of the crank are the same as other Bultaco five speeds, and showed no signs of excessive wear or failure packing around the larger crank. The connecting rod bearing and crank pin is not the same as

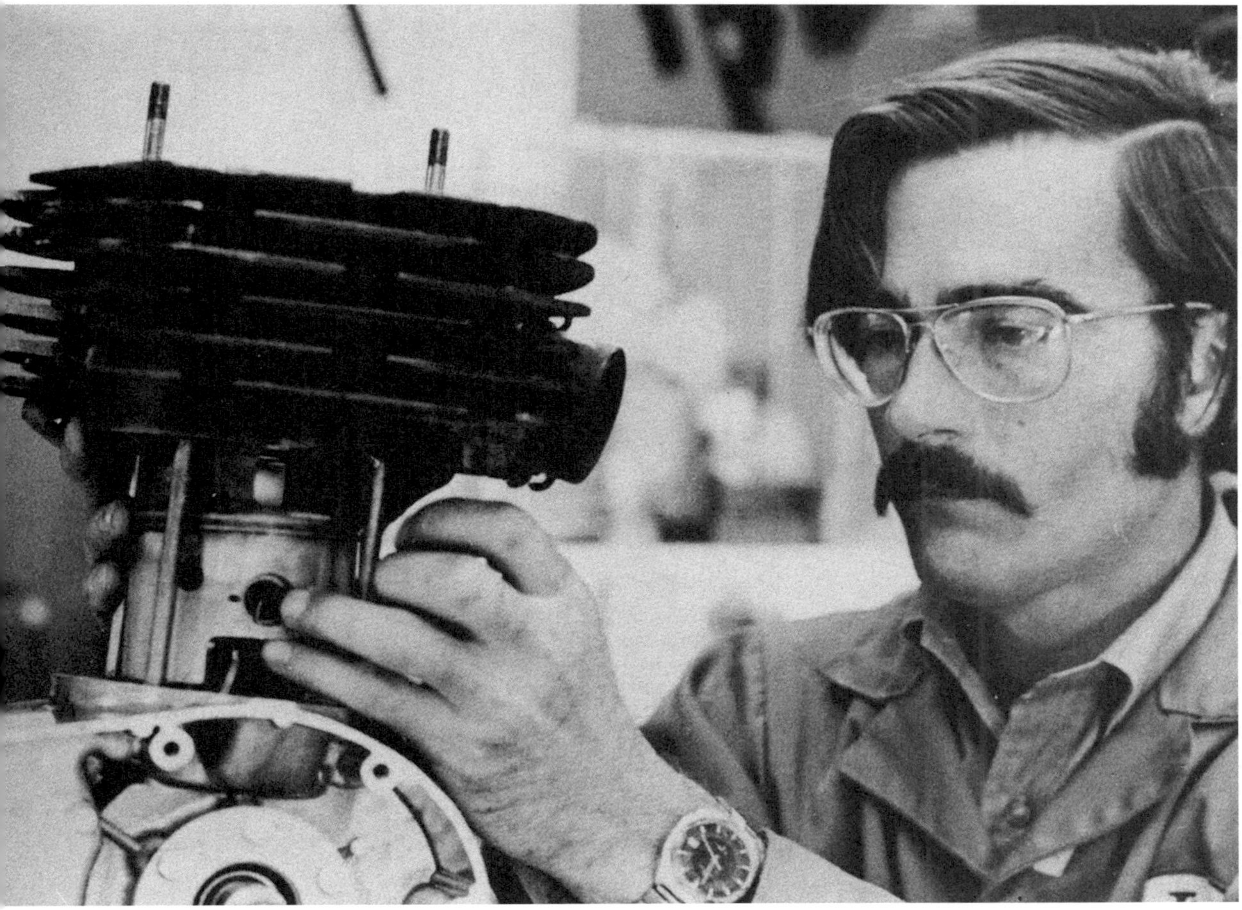

TOP— Tom Patton of Bultaco Services dismantles the 350cc Pursang to let the American public have its first look. Note large transfer port casting in the case side.

ABOVE— The head and cylinder are of the new massive finned design.

LEFT— The intake and exhaust ports are unbridged to reduce turbulance which can occur in the large displacement engine.

the smaller machines because the rod and rod bearing are both wider and of larger diameter. In a word, the only thing the same about the 250cc and the 350cc cranks are the main bearings, in spite of early grapevine information to the contrary.

The clutch is of the new all-steel TJT design and showed almost no wear of any kind. In short the engine wear was so slight that the engine could have been reassembled with nothing more in the way of parts than a new gasket set.

An additional improvement found on the 350, (and Tom tells us all the new competition Bultacos will have it) are redesigned crank seal retainers with double seals. This should keep crankcase compression up to par longer than before.

Both Rapp and Bailey reported the horsepower available from the engine was more than up to the job of wailing around in the open class. It appears that Bultaco has a good thing going. It will be interesting to see if any further changes are made on the production engine, and if it sells as fantastically as the 250 version. PC

Two transfer ports are used on each side to give the cylinder a total of five ports above the piston crown.

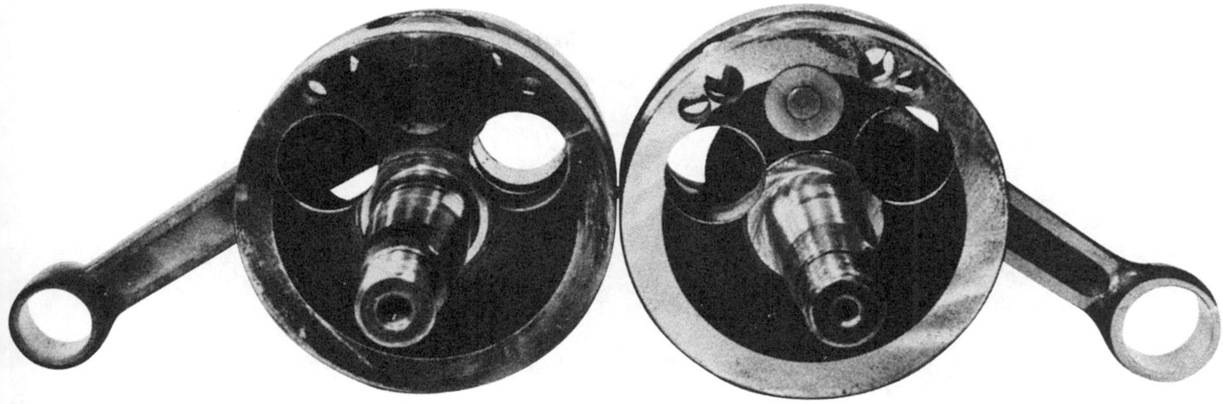

The 350cc crank (right) has a larger connecting rod and crank pin than the 250 crank shown on left.

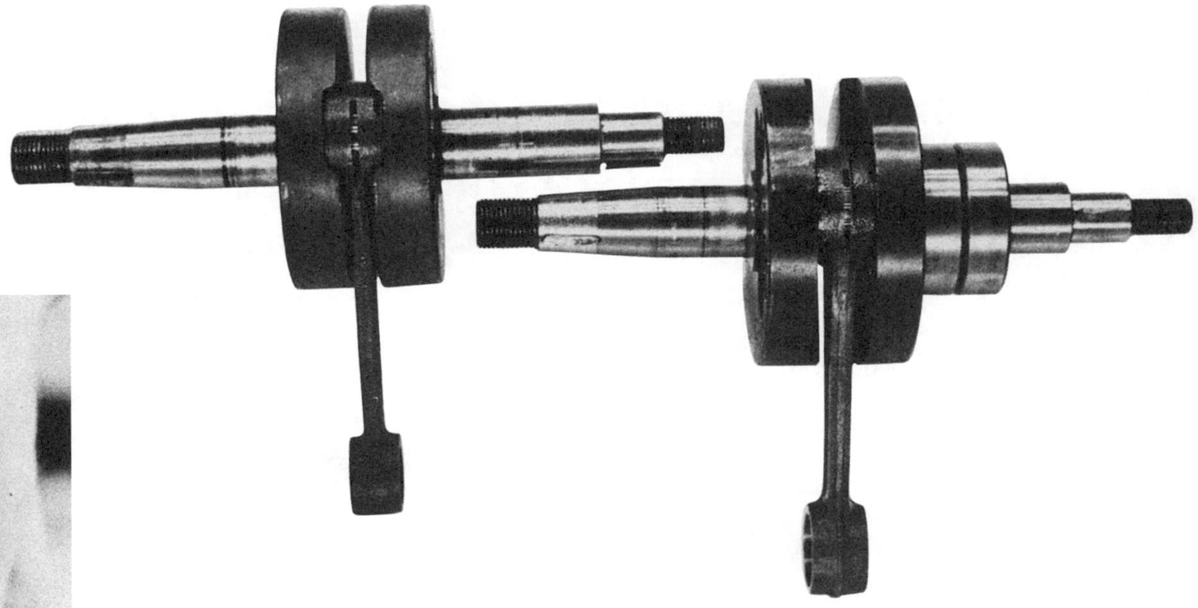

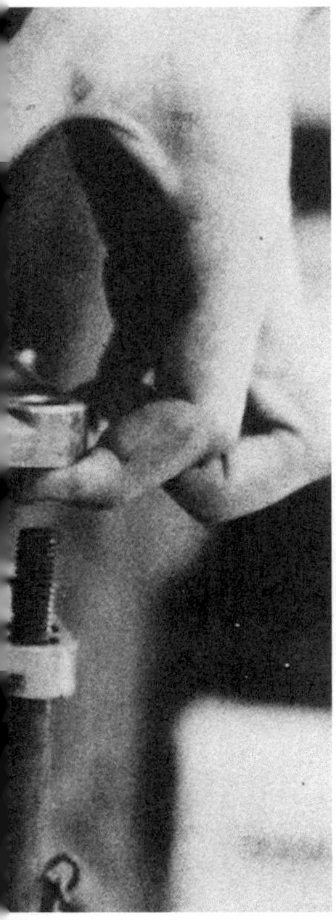

The 350 crank on the right shows larger rod bearing and entirely different cut flywheel halves. The 350cc crank flywheels are 2.5 mm larger than the 250. The lower end of the engine is entirely new and not the same as the other Bultaco five-speeds.

The cases disassemble the same as other Bultaco five speeds, but both the castings and the machining are quite different.

125cc BULTACO PURSANG MK-5

Without a doubt Senor Bulto builds motorcycles for people to have fun with. The elderly gentleman from Barcelona, Spain, has certainly made a dent in the world off-road market with his fine line of two-stroke singles. (He has also dabbled in Grand Prix road racing with some very quick 125s and 250s.)

The Bultaco factory is Spain's largest producer of motorcycles. Like most maunfacturers they began with the production of transportation-type machines and evolved into off-road bikes as their domestic market discarded the motorcycle for the automobile. About half of their total production of motorcycles comes to the U.S. in the form of enduro bikes, trials bikes, and flat out racers, all for the truly competition minded motorcyclist.

Our test machine for this issue is the exciting 125cc Pursang Mark 5 moto crosser. Truly a machine built for one purpose, winning races. Although the Mark 5 is totally new and different from previous models it still clings to many of Bulto's traditionally unique ideas. The basic one being interchangability between it and other Bultacos.

Their Mark 5 engine is built up from the same basic powerplant that does so many different jobs in the entire Bultaco line. By increasing the cylinder bore and piston size it can be a 175, 200, or even a 250. (They're even experimenting with a 325cc unit utilizing the same cases and stroke.) Everything below the cylinder is capable of withstanding greater loads than the 125 will produce which should make this bike indestructible. And if you feel like making a change up in class all it takes is another top end, carburetor, and expansion chamber, and you're competitive!

OUR STAFF EXAMINES STILL ANOTHER BARCELONA MISSILE

The 125cc Pursang engine is shrunk from the 250. It can be blown up again stopping at 175 or even 200. The machine we tested was the strongest 125cc moto-crosser we've ever ridden.

To solve the question of how many cc's are in a Bultaco just look at the rear sprocket. The bigger the sprocket the smaller the engine. This one is drilled to knock off a little weight.

Evidence of Senor Bulto's wizardry lie in the many uses he has tailored from one basic engine. With cases and castings all identical and a single 60mm stroke his current 5-speed powerplants do everything. Other than displacement changes, which he manages by altering the bore diameter (as previously stated), Bulto changes port timing and cylinder heads for desired compression ratios, carburetors, exhaust pipes, and finally, various amounts of weight added to the crankshaft assembly. The end result are engines whose personalities can be changed by simply bolting on other Bultaco pieces.

Bultaco claims 25.4 horsepower at 10,000 rpm from their robust little engine. By the seat of the pants it's the fastest 125cc MX we've ever tested. In conjunction with all that horsepower is a broad power band that is useful even on a bumpy race course. The chassis is the same for all the '72 Pursangs, and total weight of the bike is 211 pounds with an empty fuel tank.

The 56-inch wheelbase could be thought as longish for a 125cc racer, in practice it's not. The bike handles flawlessly on the moto cross race course, going where you point it and nicely soaking up the bumps with its Bultaco suspension. It tracks through a turn well, and too much throttle brings the front wheel into the air slowly. A shorter wheelbase would not be so easy to live with.

You can sit or stand on the machine with comfort as all the levers are well positioned. A narrow fuel tank and large seat are part of the package that make this one a good racer. For '72 they have discarded the old squarish fenders and replaced them with conventionally-shaped ones of the same fiberglass material.

Under the seat is a new fiberglass air filter housing that is practical and out of the way. Inside lives an urethane foam air filter that is cylindrical-shaped and capped off at one end. The air filter is held in place with a metal plate and three springs. (In the past Bultacos have been weak in their air filtration devices, but it seems this model has taken 10 giant steps forward.) The air filter box has a cover on each side that you can get to just by removing either of three screws. The oil wetted air filter is larger than necessary for a 125 so the bike could probably be raced for several meetings before the air filter would have to be serviced. And then it's just a simple matter of removing a couple of screws and three springs.

A single down tube steel pipe frame is used that is all welded together and typically Bultaco. Rubber tipped turning stops are both strong and useful in keeping a fork leg away from the gas tank. The head piece is well gussetted and uses Timken tapered bearings to support the forks. The swing arm

The new rear axle mount is boxed in for rigidity. The rugged hub is used on even the 350cc Montadero. Note the steel plate on the near side and 6mm nuts and bolts between the spokes.

This little fiberglass fuel tank holds just enough to feed the engine for one moto cross heat, about a half-hour.

fulcrum is bushed and anchored at the outermost points of the cradle. The swing arm itself is quite new this year with the rear wheel axle slots being boxed in. The result is a swing arm that is practically twistproof.

The Mark 5 anchors its rear brake backing plate with a long steel rod that doesn't serve any purpose by being so long. If it were bolted to the frame instead of the swing arm, and a bearing installed in the backing plate so it could rotate a few degrees, then the Pursang would have a floating rear brake. Maybe Bultaco is slowly working up to it year by year.

The brakes, however, were very effective for a moto cross bike and we really can't complain. Bultaco's front hub is the newest, most talked about feature of the whole Pursang line. They have made a very lightweight conical hub from steel. It looks as though they just stuck a big ol' hunk of metal in a lathe and machined themselves a complete wheel hub. The end result is a unit lighter than their small alloy hub used on the Sherpa trials bikes that, because it's made of steel, is much stronger where the spokes tie in. The backing plate and brake shoes are the same as found on the Sherpa Ts.

A huge 56-tooth rear sprocket is used that has been lightened with more than a dozen large diameter holes. A large 5/8 pitch chain is used to get the power from the engine to the rear wheel. The Pursang's gearbox has a set of close ratio gears that are correct for this engine and moto crossing.

Sometimes you have to downshift twice getting into a turn, but the shifts are smooth and never miss. An all steel clutch similar to their trials bikes is used. These plates won't fry and are capable of withstanding a lot of in-gear revving on the starting line. It has enough torque to spin the 4.00 x 18 knobby off the line in low gear, and when you catch 2nd it will carry the front wheel in the air until you have to shut it off for the first turn.

It takes horsepower to do this and it wasn't very many years ago that 25 horses were a lot for even a 250. This machine is getting 2 hp for every 10cc's. Part of the reason for this tremendous power increase can be laid to the much modified Amal concentric carburetor. They have repositioned the pilot jet in the main body to improve low speed performance. A new float bowl has been cast which produces a more constant fuel flow to the main jet and also has a higher float level. The bike doesn't surge as it goes bounding over jumps which is the habit of many concentric carburetors. Even the float tickler is a big difference. You push it down on one side to wet a cold engine and the fuel comes rushing out a hole on the other side of the body.

The machine starts super easy and you have to work at it to load it up. Most of this can be laid to the healthy spark produced by Femsatronic's capacity discharge ignition system mounted on the left side of the engine. The only moving part are the magnets cast into a flywheel, there are no contact breaker points. Once the ignition timing is set there is nothing to wear, therefore, nothing to adjust. The Femsatronic's are said to put out a healthier mid-range spark than the old pointed magnetos.

Its finish is very pleasing to the eye, plus being practical. The chassis and fork legs are painted silver. Fenders and part of the gas tank are metallic silver formed into the fiberglass. The air filter side covers and remainder of the gas tank are finished in a rich red. Black wheel hubs and polished Akront alloy rims complete a motorcycle that is destine to win a whole lot of races! ●

Bultaco forks, Akront rim, and the all new conical front hub. Although made from steel it is lighter than Bultaco's tiny trials front hub and a whole lot stronger.

Toothy brake pedal and footrests are necessary for racing in the mud. The bike uses the same large pitch chain as its bigger brothers.

MAKE BULTACO
MODEL 125cc PURSANG MK-5
PRICE AS TESTED $795.00
MAXIMUM SPEED AS GEARED 60 mph
WEIGHT 211 lbs. Dry

ENGINE

Engine Type .... SINGLE CYLINDER, 2-STROKE
Bore 51.5mm
Stroke 60mm
Displacement 124.98cc
Compression Ratio 14:1
B.H.P. at R.P.M. .... 25.4 AT 10,000
Carburetor .... 30mm AMAL CONCENTRIC
Ignition .... FEMSATRONIC C.D.I.
Starting System .... KICK
Lubricating System .... OIL MIST

FUEL AND OIL

Oil Capacity .... MIXED WITH FUEL 25:1
Oil System .... IN FUEL
Fuel Capacity .... 1.3 GALLONS
Fuel Recommended .... PREMIUM
Approximate mpg .... N.A.

FILTERING SYSTEMS

Air .... OIL WETTED FOAM
Oil .... N.A.
Fuel .... SCREEN

FRAME

Frame Type .... SINGLE DOWN TUBE
Weight Distribution ...... FRONT: 47%; REAR: 53%
Wheelbase .... 56 INCHES
Ground Clearance .... 9.5 INCHES
Peg Height .... 12 INCHES
Seat Height .... 30.5 INCHES

SUSPENSION-TYPE

Front .... TELEDRAULIC
Rear .... SWING ARM

SUSPENSION-TRAVEL

Front .... 7 INCHES
Rear .... 4 INCHES

DRIVE TRAIN

Clutch Type .... MULTI-DISC, ALL STEEL
Primary Drive .... DUPLEX CHAIN
Countershaft Sprocket .... 11 TEETH
Final Sprocket .... 56 TEETH
Final Drive .... SINGLE ROW CHAIN
Gear Ratios:
1st .... .376:1
2nd .... .513:1
3rd .... .670:1
4th .... .832:1
5th 1.100:1

TIRES AND WHEELS

Front .... 2.75 x 21 PIRELLI KNOBBY
Rear .... 4.00 x 18 PIRELLI KNOBBY

IMPRESSIONS	Poor	Good	Excellent
Throttle Response			●
Acceleration			●
Power Band			●
Starting		●	
Engine Noise		●	
Muffling	DOES NOT APPLY		
Vibration			●
Handling			●
Choice of Tires			●
Suspension			●
Rider Comfort			●
Transmission			●
Instrumentation	DOES NOT APPLY		
Lighting	DOES NOT APPLY		
Toolkit	DOES NOT APPLY		
Paint and Chrome		●	
Styling			●
Mileage		●	
Braking		●	

A few years back Bultaco entered the big bore racing market with the 350cc Bandido. At that time one of the West Coast's top moto-cross riders rode the bike one lap around a local track and then parked it saying, "There's not a rider in the world that can win a race on that machine." That statement was not quite the truth, but at best the Bandido was almost impossible to ride. It was terribly pipey and didn't handle. In fact there wasn't too much it did well on a moto-cross course. On smooth tracks it was a blast to ride because it did have power, and like most 'Tacos it liked to slide.

Came 1972 and rumors started to fly about an all-new Bultaco for the big bore class. And the rumors were true. Bultaco indeed had a model for the 500 class in moto-cross racing. But to the surprise of many, including us here at POPULAR CYCLING, the displacement was only 326cc, a bit odd when you realize that just about everyone is trying to build two-strokes as close to the 500cc mark as possible.

The reason various manufacturers keep boring out their moto-cross machines is to get added torque, the only real difference between a good 250 and a 500cc cycle. Somewhat surprisingly, but happily, we can report that the 326, which is actually called a 350 Pursang, has all the low-end torque that any bike needs. With stock gearing, which is good for around 70 mph, starts on a normal moto-cross track have to be made in second gear, first does nothing but dig a big trench.

One of the big reasons for the outstanding performance of the 350 Pursang can be traced to the weight, a grand total of 218 pounds dry, which means a race weight of under 230 pounds, extremely light for a bike in the 500 class. This light weight is an advantage in another way to the rider. Physical force needed to change directions is less with a light machine, so, the rider of a 350 Pursang shouldn't get as tired as his competitors who are man-handling bikes that might weigh as much as 30 pounds more than his own steed. Every little thing helps in a 30-minute moto.

Down through the years, Bultaco has always built motorcycles that produced power. Without resorting to reed valves, rotary valves, or any type of valves or trick induction, Bultaco cycles have been equal in power or in many cases produced more horsepower than their competitors. The reasons behind this are very simple. Bultaco has taken what is a good engine and refined it through the years to where now it's one of the most reliable and potent power-plants available. The 350 is somewhat different inside from the 250; larger flywheels, bigger wrist pin and big end, but for all practical purposes the basic design is the same as the well-proven 125, 175, 200 and 250cc Bul engines.

The mixture is provided by a 36-mm Spanish Amal Concentric carburetor that ran surprisingly clean throughout the rev range. The Concentric carburetor isn't our choice for the best-engineered carb in the whole world, but the slow-speed circuit has been redesigned in this jug and the bike pulls well when the engine revs are down in the lower portion of the rev range. In fact, the lower end torque is one of the surprising features of the 350 Pursang. Before we rode the machine we expected that perhaps the 350 would run on the top end with various 400cc moto-cross machines, but we didn't expect it to have the low-end muscle that is so necessary. Much to our amazement the 350 came out of the hole and off slow corners like any 400. In fact, there's a few 500cc's around that won't get from corner to corner as fast as the 350. Again much of this can be traced to the excellent power band and the light weight. It's been said by some very knowledgeable European racing engineers that eight pounds equals one horse-power. If this is true, and we see no reason not to believe them, then the 350 Pursang enjoys quite an advantage

LIGHT OF FOOT AND HEAVY ON HORSEPOWER

Bultaco 350 Pursang

TOP— The overall view shows that the 350 doesn't look a great deal different from the 250 Pursang. About the only noticeable difference is the size of the carburetor, yellow side number plates, and the dual plug head.

ABOVE— A cast iron hub is used to keep down the weight and provide extra strength for the spoke mountings.

Bultaco 350 Pursang

Both the chain guide and guard are functional, a change from some machines where they're too far from the chain to do any good. An indexed snail cam is used for chain adjustment.

This nifty little side stand fits on either side of the bike. Footpegs are of the self-cleaning type.

Both plugs fire at the same time to insure a good flame front.

Akront rims and Betor forks, with seven inches of travel, are features at the front of the bike.

The 36mm Amal does a surprisingly good job of feeding the hungry 326cc engine. Four springs hold on the exhaust pipe.

A skid plate is a definite must if the bike is going to be ridden in the desert. One rock could clean out the whole underside of the engine.

Without actual testing with a sound meter we'd say that this is the best muffled motocrosser available for sale.

over the majority of its competition before the race even starts.

In the past, Bultaco shocks have been replaced by many 'Taco riders. Their dampening hasn't been the best and they've usually been oversprung, a characteristic found on most Japanese motorcycles. The shocks used on the 350, five-way adjustable Betors, seem to be just right for the weight of the bike and average rider. There's probably some riders who will want to change to another brand of shocks but at least we're convinced that the Betors are good units.

The Betor front forks are as good as you can buy. More than one international moto-cross star has publicly gone on record and stated that the Betor units were the best forks available. They have seven inches of travel and really fantastic dampening. No fork brace is used and none is needed. The legs stay parallel no matter what stress they're subjected to.

The frame is a single downtube unit that seems quite free from flex. The basic design of the frame and head angle tends to make us believe the bike is meant more for fast moto-cross courses and the desert instead of the tighter-cornered moto-cross courses found in Europe and on the Eastern Seaboard. In tight corners the tendency is for the front end to plow somewhat. This situation can be easily corrected by gassing it and letting the over-abundance of power swing the rear end around. This heaviness, as it's sometimes called, in the front end is one that will make the 350 a good handler in the desert. A bike that handles spot-on in moto-cross usually handles too quickly in the desert. All it takes to cure this is the addition of a degree or two in the fork angle, the 350 Pursang already has it.

The brakes on the test bike weren't up to Bultaco standards and we suspect that they were improperly seated into the cast iron hubs. Instead of casting up the brake hub in aluminum, and then inserting iron drums for the shoes to wear against, Bultaco has opted for a complete cast iron hub. It's lighter than the more conventional arrangement and there's greater strength where the spokes are attached to the hub. It's a neat setup.

One thing we definitely didn't like about the 350 was the seat. If you weren't sitting right up on the gas tank at all times, the hump where the fender passed under the saddle got you right in the tailbone. It wouldn't be too bad during a 20-minute moto but for a four- or five-hour desert go, padding definitely has to be added.

We commend Bultaco on their silencer arrangement. The bike comes standard with a muffler that really works. We don't have a sound meter (we DO plan on getting one) but listening to the bike we'd say that it is more quiet than many 250cc street/trailer combination machines. The muffler cuts some power from the top end but returns it in the form of mid-range torque, a better-than-even trade in our estimation.

All in all we were very favorably impressed with the 350 Bultaco Pursang. It ran hard and long and was a true joy to ride. It was one bike we were genuinely reluctant to return. **PC**

BULTACO 350cc PURSANG

PRICE
$1195.00

ENGINE

ENGINE TYPE
five-port, two-stroke single
HORSEPOWER @ RPM
N.A. @ 7200
TORQUE
N.A. @ 6000
BORE AND STROKE
83.2 by 60mm
DISPLACEMENT
326cc
COMPRESSION RATIO
10.5:1
CARBURETION
36mm Amal Concentric
IGNITION
Femsatronic dual CDI

TRANSMISSION

SPEEDS
five
CLUTCH
all steel, wet multi-plate

CHASSIS

LENGTH OVERALL
82.6 in.
WHEELBASE
55.9 in.
GROUND CLEARANCE
9.6 in.
WEIGHT
218 lbs. dry
FRAME TYPE
single downtube
TIRE SIZE
front: 3.00–21
rear: 4.00–18
BRAKES
front: single-leading shoe
rear: single-leading shoe

POPULAR CYCLING IMPRESSION

BULTACO TIRON

The littlest Bul' kicks up quite a fuss

One of the more popular sized machines today is the mini motorcycle for the children. It seems that almost every manufacturer is building a bike for the youngsters: Honda, Yamaha, Suzuki, Montesa, and Bridgestone are just some of the brands that come to mind. Now Bultaco has introduced one, and it just might be the best of the group. It's called the Tiron, which means tyro, or novice, in Spanish, but the bike is more for the experienced child instead of the rank beginner.

For openers the bike is propelled by a 100cc Bultaco mill that's much the same as a Lobito engine. It's a five speeder that really gets it on. The top speed is well over 50 mph and as such we don't advocate a first-time rider picking this bike for his first one. It's a lot of mini motorcycle.

One thing it has that we enjoyed was the full hydraulic suspension, both front and rear. The forks have between three and four inches of travel and are dampened as well as full-sized 'Taco forks. The rear shocks work better than any shocks we've come across on the mini machines. And well they should because the Tiron carries the heaviest price tag of any of the ultra lightweights, $575.

The bike is basically set up as a trials machine. The thinking behind this from the factory is that trials are not a speed contest, but yet trials will do more to teach children throttle control than any other form of the sport. Throttle control is the secret to any form of riding, trials just take more control than any other aspect.

The machine seems to handle better when the rider is standing, more or less like a full-sized trials bike. However it handles quite well when the rider is sitting down. The bike has more than enough power to powerslide around corners but because it is short coupled we suggest you take it fairly easy until you're used to how quickly things can happen.

An important feature on any machine a man would put his child on are the brakes. We're happy to report those on the Tiron work extremely well. They're very powerful but locking up the wheels doesn't seem to be a problem. An adult only needs two fingers for the front brake but it's probably best for a child to use all four. The rear stopper is a foot pedal like a full-sized machine and it takes a great deal of stomping to lock up the wheel. That's good because too many kids could get in trouble if the wheel was easily locked up. Both brakes are cable operated.

The Tiron has many of the features

Like the bigger 'Tacos the Tiron has a secondary chain tensioner.

The powerplant of the Tiron is a full-sized 100, the same as found in the 100 Lobito. The only difference is the carburetor size. It is also possible to put up to a 250cc barrel on these cases.

To keep the engine mild for the youngsters a very small carburetor is used. The bike will still top out at well over 50 mph.

found on the larger models in the Bul' line. There's a secondary chain tensioner to keep chain snatch to a minimum, a chain guard is provided, there's full lighting equipment, and a small skid plate. Tires are of a universal tread, not the best for either street or serious off-roading, but they're a good compromise. The bike is quite light, 152 pounds, and this compares favorably with its competition. The footpegs fold and are made from the same sturdy stock as that of the bigger Bultacos.

Gas mileage won't be any problem because the Amal Concentric usually found on the 100 Bultaco engine has been replaced by a very small Zenith. We'd estimate the mileage will be well over 100 mpg. The air cleaner fits directly on the carburetor mouth.

The frame is a single downtube unit that can't help but be rigid if for no other reason than it's small. There's not enough of a span anywhere long enough to allow flexing to be a problem. Even then there are gussets at all the stress areas; maybe they are there because it is possible to put a 250cc barrel on the lower end. What an animal that would be.

The fiberglass work is outstanding. The gas tank and rear panels are a one piece unit and the finish is flawless. The seat, while small, is comfortable and a child should be able to go all day without any complaints.

About the only thing we didn't like was that the exhaust pipe dumped out right at the left shock. In just a very few miles of riding the shock was covered with oil passed through the muffler, which by the way is pleasantly quiet. Bultaco plans on importing very few Tirons; hopefully they'll redirect the muffler outlet and this one problem will disappear.

Actually there's one other problem with the Tiron. It's kind of like electric trains. A father will buy this bike for his son, and chances are very good the son will have to take a stick and beat his dad to keep him away from it. The Tiron is a bike that the whole family can enjoy. **PC**

Forks are hydraulic and dampen surprisingly well.

The lighting is adequate and the single leading shoe front brake will definitely pull the speed down when the little tyke pulls the lever.

TEST: THE 1973 BULTACO ALPINA

A SECOND LOOK AT THE MACHINE MODERN CYCLE UNVEILED A YEAR AGO.

Just one year ago *Modern Cycle* was privileged to be the first motorcycle publication to test the newly-created Bultaco Alpina. We praised it highly, and justly so because you see, where there is a need, and that need is fulfilled, somebody did something right.

The Alpina is that something that filled a need. For years people have been purchasing Sherpa T trials machines just to pleasure ride those tight tricky trails found way out in the forests and deep in the woods. To a man they all would have liked larger fuel tanks and softer seats. Many would prefer a different gear selection than what is found on the T's. But change the chassis? Alter the broad torque curve of that smooth 250 engine? Never!

Basically the Alpina is a Sherpa T. Senor Bulto sought the services of Sammy Miller to help him make a trials machine, the *best* trials machine. From 1965 till now the Sherpa T has been gradually changed and improved to the point of perfection. The Sherpa T chassis gets its power to the ground like no other, and handles extremely well under all conditions. These changes had been gradual, like every time a new batch of bikes was to run down

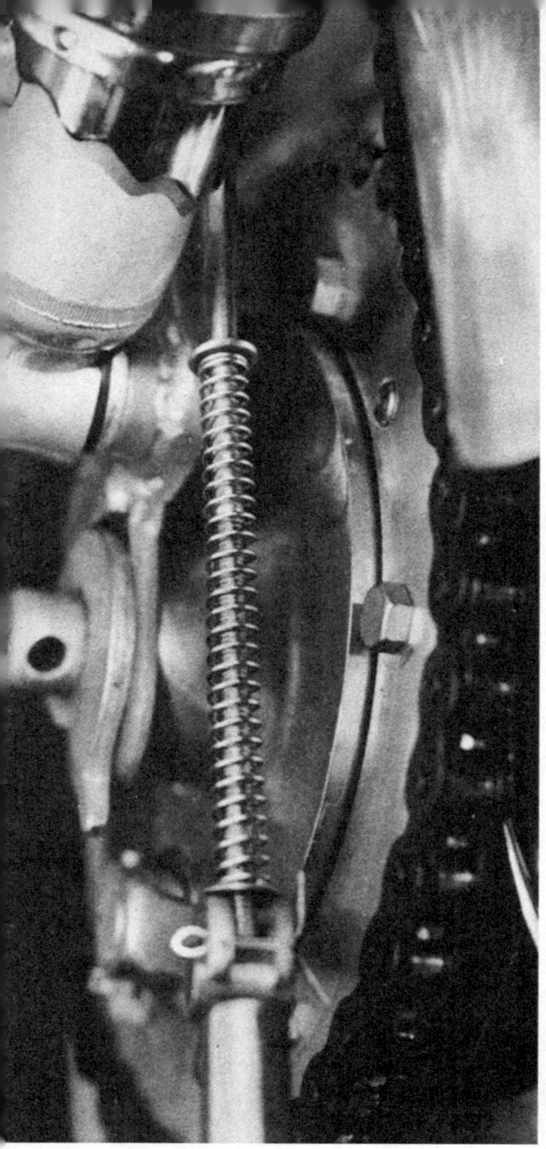

BULTACO ALPINA

the assembly line there would be something different; something better, an improvement. The Alpina took advantage of all this by utilizing the Sherpa T chassis and fulfilled the needs of the long distance trail rider.

The 1.5-gallon T gas tank was increased to 2.6 without becoming excessively wide (they added a cavity down and back near the front of the seat). The seat itself was made larger and more comfortable than the boards found on trials bikes. The Alpina mounted its front fender high to suit U.S. tastes and also a speedometer. The trials gearbox with its ridiculously close first to second gear placement was swapped for that of the Matador, an evenly spaced, wide ratio, five-speed set. (Trials work requires the lower gears closer together because of the preciseness with which the machine must perform over low speed, difficult terrain.)

Although we were thoroughly impressed with the Alpina, we did come up with some well founded faults on that particular bike. It was fitted with trials shocks that proved inadequate for fast trail riding. The seat was still a bit firm. Some thought the trials bars were fine and some didn't. So, like the Sherpa T, the Alpina has gone through a minor change each time a new batch was built. All totalled, there are roughly 40 improvements since the first Bultaco Alpina we tested just last year.

We'll list the more impressive ones: You can immediately spot the thicker padded seat that came out a few months ago along with the motocross handlebars (I personally prefer the trials-type). The fuel tank is a couple of layers thicker and the odometer is now a tenth trip model. (We made comment on that.)

Gone are the Spanish Amal steel levers, they have been replaced with alloy forgings. The front fender, although still mounted up under the triple clamp, is hung at three points off a single bracket, that makes it lighter and less complicated. The engine had three plastic inspection plugs in its cases, the primary filler plug is now aluminum. (Gradually the others will be changed, too.)

The frame gussets that mount the swing arm pivot and footrests have been moved outward. Footrests now pin onto welded-on

stubs instead of bolted-on plates, and the fulcrum nuts are recessed. The chain tensioner has now been turned around and is a leading link instead of trailing. The ½-inch pitch rear chain has been changed to the stronger ⅝ths. The rear backing plate is a closer fit to the hub so less crud can find its way in. They changed to full-size rear shocks a long time ago, but the plastic bumpers on the fork stops are new.

Removing the tank/seat combination was always a pain because invariably you would have to hunt for a lost nut or two. These nuts are now fastened to the frame. A stoplight switch has been added and a resistor within the stoplight circuitry. If the filament goes the motor still runs.

The engine has had its little improvements, too, like double seals on the crankshaft.

Finally, a chain oiler has been built into the swing arm. Only one problem, it's gonna be tricky trying to get oil into the oiler, you need a right angle screwdriver just to remove the filler plug. Oh yes, the rear sprocket is an alloy one.

We would like to see them mount the fender next to the front tire like the Sherpa T. After all, the benefits gained from better engine cooling far offset the reducton in unsprung weight found with high fenders. Those big, wide motocross bars are dumb on this bike because it is not comfortable for stand up riding. The trials bars would be much more practical especially with a chassis that is designed for off-your-arse riding. It still has a slow turn throttle, but like the bars, that's another matter of personal preference.

Lastly, and we spoke of this a whole year ago, the darn speedo should be driven off the front wheel, not the rear. It's really a shame to pick on such a fine motorcycle. The Alpina, with its utterly different look, is one of the more totally successful motorcycles around today.

It is light, agile, and quick handling. The engine doesn't come-on with a surge, it just pulls. The bike works in such a manner that it gets there quickly, not fast. You can do silly things like sliding and wheelies, but it's got to be taken into the woods to be thoroughly enjoyed.

The Alpina will submerge to its air filter that is just under the seat and pull up a muddy or grassy bank without breaking traction. It has a way of picking a trail up or down a rock strewn mountainside and scoots up a dry riverbed with ease.

With an engine that's not the least bit ornery, a gearbox without holes, and a chassis designed to get the utmost from weight transfer and traction, the Alpina doesn't have a peer.

In true trials tradition you get a cummerbund with each Alpina. Actually, it is a tool pouch you buckle around your waist so you can carry things along, like your very own rock to land on. In mountain riding the Alpina is good for 70 to 90 miles before the 2.6-gallon tank runs dry. Its sheer ease of handling makes it untiring to ride, and the very quiet sound emitted from the second muffler doesn't hammer at your eardrums.

Alpinas are not fitted with batteries. Their lighting system comes from a flywheel magneto that also supplies spark to the plug. Although most Bultacos use the more modern pointless CDI-type ignitions, Sherpa Ts and Alpinas retain the pointed-type. The reason for this is the spark at low speed running occurs at a more predictable time and is stronger, very important for trials riding.

The lighting coil puts out AC current to a 6-volt headlight and taillight. A fairly bright beam is thrown out by the headlamp that is ample for afterdark chugging. A switch on the left bar handles both beams, off, and has an engine kill button. There is no horn as the machine is not intended to be sold as street legal, although it will pass in many states with just the addition of a bulb-type horn.

The stoplight gets its power off the other end of the primary ignition coil. This means as long as the engine is running there is power to the bulb independent of the head/taillight switch. Problem was when the headlight filament would part there would be no more spark to the plug. All they needed to correct this was to put a resistor in the stoplight circuit, and this they did with the '73 Alpina.

It seems there is little left to do to improve the bike, but knowing the way things are done at the Barcelona works, the Alpina, like the Sherpa T, will continue to get better. We'll have another look about this time in 1974 and see. •

MAKE BULTACO
MODEL 250cc ALPINA
PRICE AS TESTED $975.00
MAXIMUM SPEED AS GEARED 72 mph
WEIGHT 205 lbs. DRY: Fuel Tank Empty

Distributor:

East: CEMOTO EAST
P.O. Box 1065 Schenectady, New York 12301

West: BULTACO AMERICAN LTD.
P.O. Box 101 Santa Clara, Calif. 95052

ENGINE

Engine Type	SINGLE CYLINDER, 2-STROKE
Bore	72mm
Stroke	60mm
Displacement	244.3cc
Compression Rato	9:1
B.H.P. at R.P.M.	20 AT 5,500
Carbueto	27mm AMAL CONCENTRIC
Ignition	FEMSATRONIC, POINTED
Starting System	FOLDING KICK
Lubricating System	OIL MIST

FUEL AND OIL

Oil Capacity	IN FUEL
Oil Recommended	BARDAHL VBA 50W
Fuel Capacity	2.6 GALLONS
Fuel Recommended	PREMIUM
Approximate mpg	45

FILTERING SYSTEMS

Air	OIL WETTED FOAM
Oil	N.A.
Fuel	SCREEN

FRAME

Frame Type	SINGLE DOWN TUBE
Weight Distribution	FRONT: 48%, REAR: 52%
Wheelbase	52.5 TO 53.5 INCHES
Ground Clearance	11 INCHES
Peg Height	13.8 INCHES
Seat Height	32 INCHES
Handlebar Width	34 INCHES

SUSPENSION-TYPE

Front	TELESCOPIC FORK
Rear	SWING ARM

SUSPENSION-TRAVEL

Front	BETOR 7 INCHES
Rear	BETOR 4 INCHES

DRIVE TRAIN

Clutch Type	MULTI-DISC, WET
Primary Drive	SINGLE ROW CHAIN (2.375:1)
Countershaft Sprocket	14 TEETH
Final Sprocket	52 TEETH
Final Drive	SINGLE ROW CHAIN (3.71:1)

Gear Ratios	Internal	Overall
1st	.287:1	30.415:1
2nd	.442:1	19.886:1
3rd	.625:1	14.064:1
4th	.821:1	10.706:1
5th	1.000:1	8.790:1

TIRES AND WHEELS

Front	2.75x21 FIRESTONE KNOBBY
Rear	4.00x18 PIRELLI KNOBBY

IMPRESSIONS	Poor	Good	Excellent
Throttle Response			●
Acceleration		●	
Power Band		●	
Starting			●
Engine Noise		●	
Muffling		86Db(A)	
Vibration			●
Handling			●
Choice of Tires			●
Suspension			●
Rider Comfort		●	
Transmission			●
Instrumentation		●	
Lighting		●	
Toolkit		●	
Paint and Chrome		●	
Styling			●
Mileage		●	
Braking		●	

BULTACO 175 & 350 ALPINAS

Don't Think Just Enduro Bike. The Lightweight Alpina Concept Is Something New And Unique — Borrowing From Both Enduro And Trials. Call It "The Great American One-Day Machine."

Cycle World Road Test

■ BULTACO'S Sherpa T, the world reknowned observed trials machine developed by many times world champion Sammy Miller, has won over a large following in the United States and in other countries as well. The Bultaco Matador, on the other hand, was developed by the factory for competition in the famed International Six Days Trial and is a more ruggedly built and heavier machine than the Sherpa T, with long distance enduros being its strong point.

It soon became evident to Bultaco representatives in America that there was a demand for a Sherpa T which would be more suitable for the popular one-day enduro contests in this country. Although the Matador was a machine that could be used for these events, many riders wanted something lighter and more maneuverable for use in these short events as well as a mount for strict pleasure riding.

The Alpina is just such a machine. However, riding one requires a definite readjustment of riding style. Why? Because this ultra lightweight will not slide due to a considerable rear weight bias. And that's not all. Alpinas really feel strange for the first few miles, unless of course you have been preconditioned to the trials attributes of high pegs and low seat ala Sherpa T.

Actually in its first guise the Alpina held many of the virtues of the Sherpa T: good bottom end power, a narrow profile and a redesigned seat/gas tank combination with greater fuel capacity and a softer padded seat. In its present form, the Alpina is almost 50 lb. lighter than the more fully equipped Matador and is the ideal set up for short enduros and pleasure riding. But the first Alpina, introduced near the end of the 1971 season, was available with only a 250 engine, which was okay if your bag is to ride the 250 class in enduros.

For 1973, Alpinas are available in four of the most popular engine sizes for enduro competition: 125, 175, 250 and 350cc and almost all the cycle parts are the same. The same crankcase assemblies, except for flywheel weights, are used and the big differences to the engines are different size cylinders and pistons.

In order to get a good sampling of the four Alpinas, we requested a 175 and a 350 for test, and although the handling qualities are practically the same, the 350 has much more tractive power than the 175.

In typical Bultaco fashion, the engines feature vertically split crankcase assemblies. On the Alpinas a Femsa flywheel magneto with lighting coils is installed on the left hand end of the crankshaft while additional flywheels are mounted on the right hand (primary drive) end of the shafts. A single row primary chain transmits the crankshaft's power to a clutch with all steel plates which will take an extrodinary amount of abuse before it gets hot enough to slip or drag.

All four Alpinas have the same internal gearbox ratios which are ideally suited for the machines' intended usage. The first three gears are more widely spaced than on a Sherpa T as are the top two speeds, so keeping the engines in their powerbands is easy. >

ALPINAS

All-out performance is not a criterion for these machines so their port timing is on the mild side with good power available from ultra-low rpm right up the scale to peak power speed. The 125 is the buzziest engine in the group, developing its maximum power at 8000 rpm. The 350, on the other hand, puts out its maximum of 21 bhp at 5000 rpm and a whopping 21.5 lb.-ft. of torque at 4500 rpm. Riding the 350 Alpina is more like riding an old 500cc four-stroke Single than a medium displacement two-stroke.

The four Alpinas all use the same frame, which is practically a carbon-copy of the Sherpa T frame. A single toptube runs from the steering head back under the gas tank/seat combination to a point below the seat where it divides and continues rearward, eventually terminating in a loop which serves as an additional seat support and a mounting point for the polished aluminum rear fender.

In an effort to keep the machine's weight down the lower frame tube attaches to the bottom of the steering head and runs downward to a point in front of the engine. Here it divides into two smaller tubes which continue rearward to a point near the rear of the engine where they slope upward, forming an attachment point for the top rear suspension mounts and join with the top frame tube into the loop section.

Liberal gussetting is provided in the swinging arm pivot point area assuring minimal flex in this critical area which is so important to the handling qualities of a motorcycle.

A perforated steel bash plate is welded between the two lower frame tubes under the engine to protect the crankcase area from rocks and logs while traversing rough country.

The one-piece fiberglass gas tank/seat combination is very similar to the one used on Sherpa T models, but gas capacity is up from 1.3 to 2.2 U.S. gallons and the seat is slightly wider and more thickly padded for comfort. Here you can see the Alpina's orientation to the enduro or primarily recreational rider, who spends much of his time sitting down. The larger fuel reserve makes the Alpinas capable of running the 40 to 60-mile leg to the typical noon gas check.

Finely finished and detailed, the gas tank/seat combination may be removed by unfastening three screws for access to the ignition high tension coil, electrical connections and the air filter. The filter is located high, just below the seat, which will prohibit the ingress of mud and water during stream crossings. However, the filter material is a plastic gauze which is much too coarse to stop fine dust from entering the engine. One of the popular closely meshed foam elements available today makes a much better filter, and will be fitted shortly.

The Alpina 175, with its practically "square" bore/stroke ratio of 60.9 x 60mm, pulls surprisingly well from low rpm. Quite a bit of de-tuning has been done to this engine, which was originally fitted to the Bultaco TSS 175 road racer and put out more power per cubic inch displacement than any other Bultaco engine.

Flooding the carburetor liberally makes starting the 175 a two or three kick affiar, but the kick starter lever hangs up on the brake pedal near the bottom of its stroke and must be returned to its original position by hand. Once warmed up the engine pulls smoothly and strongly throughout its speed range with no hesitation or "portiness."

The 350, on the other hand, is more difficult to start because of its high primary compression and the fact that its kickstarter also hangs up on the brake pedal. Once running, the 350 can best be likened to a "tractor" so abundant is the power. Hills which you would think should be climbed in second gear are easily climbed in third and the engine may be lugged down to an idle without stalling.

Besides the power characteristics of the engines, the >

Lightweight aluminum alloy chainguard helps keep wear-causing debris off chain. Hard rubber chain guide is spring loaded, helps keep chain running true between sprockets even when it is too loose.

Rear-wheel drive may introduce some error into speedometer mileage readings. Note snail cam adjuster for rear axle.

Large, readable speedometer has a resettable tripmeter, especially useful to the enduro rider.

Fork stanchion nut has a flapper valve to ease the entry and exit of air while the machine is traversing rough terrain; results in a smoother ride.

constant mesh transmissions deserve the highest praise. Their short throw shift levers provide slick, positive shifts and the upper transmission ratios are so close that clutchless shifting is a breeze. With shafts supported by hefty ball bearings, Bultaco transmissions are among the strongest and most trouble-free.

The Alpina's riding position is exactly the same with a pleasant (once you get used to it) relationship between the seat, foot pegs and the handlebars. One staff member stated that he would rather have slightly higher handlebars to keep from having to arch his back while standing up, but the rest of the staffers thought the general control layout was just fine, with one exception. The rear brake levers are simply too high in relation to the foot pegs, making it necessary to raise one's foot to apply the rear brake. And, once it has been applied, great care must be exercised to keep from locking up the rear wheel.

The front brake, on the other hand, provides good stopping with a moderate amount of lever pressure. The front wheel hubs are very light in that they have a steel film applied by an electrolytic process to the inside of the drum rather than a heavier cast-in liner.

Every attempt to save weight has been made on the Alpinas. Akront aluminum alloy wheel rims are fitted to the small wheel hubs, light alloy handlebar levers are used and the marginal lighting equipment, which will be removed by most owners, is also very light. An aluminum alloy rear chain guard is fitted and a spring-loaded rear chain tensioner helps keeping the chain from being knocked off the sprockets by branches.

Another feature we particularly liked was the aluminum alloy front fender which is mounted beneath the lower fork triple clamp with a steel strap which permits it to flex, dislodging mud which may build up under the fender.

Bultaco has seen fit to use the standard silencer which is a rectangular affair with a perforated tube running through a fiberglass packing. The outlet end is significantly smaller than the inlet end, but the sound emitted is somewhat louder than we like. Additional supplementary mufflers are supplied with each machine which cuts the noise level down considerably. We rode the bikes with the additional mufflers installed and didn't notice any reduction in performance.

One area in which Bultaco could improve silencing is at the intake side. At moderate to large throttle openings a distracting roar comes from the air cleaner under the seat. It certainly isn't loud, but it is unpleasant in tone.

Handling characteristics of the Alpinas are first rate. The front forks are manufactured to Bultaco specifications by Betor and with an improved air valve in the top nut, movement of the legs up and down over the bumps is smooth and progressive. Almost 6.5 in. of travel are offered by the front forks and the five-way adjustable springs on the rear suspension units will compensate for the weight of practically any rider. These rear suspension units allow the rear wheel to follow undulations of the ground and provide nearly 4 in. of travel.

The pride in ownership of a motorcycle, be it for the street or for strictly off-road use, is very important to most owners. Detailing on the new Alpinas is better than that we've seen on previous Bultacos and is almost as good as on any machine. Aluminum parts are highly polished or smoothly painted, the fiberglass work is as smooth as any we've seen and the appearance of aircraft-type self-locking nuts add an air of professionalism to these competition-bred machines.

Of course, we predict that as many of the Alpinas sold in the United States will be used strictly for pleasure riding as for serious enduro competition. But for whichever purpose, Alpinas are very desirable machines.

BULTACO ALPINAS

175

SPECIFICATIONS	
List price	$921
Suspension, front	telescopic fork
Suspension, rear	swinging arm
Tire, front	3.00-21
Tire, rear	4.00-18
Engine, type	two-stroke Single
Bore x stroke, in., mm	2.39 x 2.36, 60.9 x 60
Piston displacement, cu. in., cc	10.7, 174.8
Compression ratio	10:1 (uncorrected)
Claimed bhp @ rpm	14 @ 5500
Claimed torque @ rpm lb.-ft.	15.1 @ 5000
Piston speed @ rpm ft./min.	2170 @ 5500
Carburetion	25mm Amal concentric
Ignition	flywheel magneto
Oil system	oil mist
Oil capacity, pt.	oil in fuel
Fuel capacity, U.S. gal.	2.2
Recommended fuel	premium
Starting system	kick, folding crank
Air filtration	oil-wetted gauze

POWER TRANSMISSION	
Clutch	multi-disc, wet
Primary drive	single-row chain
Final drive	chain
Gear ratios, overall:1	
5th	8.29
4th	10.08
3rd	13.25
2nd	18.71
1st	28.81

DIMENSIONS	
Wheelbase, in.	52
Seat height, in.	31.5
Seat width, in.	7
Handlebar width, in.	34
Footpeg height, in.	13.5
Ground clearance, in.	11
Curb weight (w/half-tank fuel), lb.	215
Weight bias, front/rear, percent	45/55

350

SPECIFICATIONS	
List price	$1127
Suspension, front	telescopic fork
Suspension, rear	swinging arm
Tire, front	3.00-21
Tire, rear	4.00-18
Engine, type	two-stroke Single
Bore x stroke, in., mm	3.27 x 2.36, 83.2 x 60
Piston displacement, cu. in., cc	19.9, 326
Compression ratio	9:1 (uncorrected)
Claimed bhp @ rpm	21 @ 5000
Claimed torque @ rpm lb.-ft.	21.5 @ 4500
Piston speed @ rpm ft./min.	1960 @ 5000
Carburetion	27mm Amal concentric
Ignition	flywheel magneto
Oil system	oil mist
Oil capacity, pt.	oil in fuel
Fuel capacity, U.S. gal.	2.2
Recommended fuel	premium
Starting system	kick, folding crank
Air filtration	oil-wetted gauze

POWER TRANSMISSION	
Clutch	multi-disc, wet
Primary drive	single-row chain
Final drive	single-row chain
Gear ratios, overall:1	
5th	7.65
4th	9.41
3rd	12.25
2nd	17.31
1st	26.32

DIMENSIONS	
Wheelbase, in.	52
Seat height, in.	31.5
Seat width, in.	7
Handlebar width, in.	34
Footpeg height, in.	13.5
Ground clearance, in.	11
Curb weight (w/half-tank fuel), lb.	224
Weight bias, front/rear, percent	45/55

It will take a better man than I Gunga Din

or THE 400 BULTACO'S TEST OF PAUL

I THINK that every competition rider — particularly if he is second-rate — nurtures the idea that if only HE had the machinery that the "works" men ride then it would only be a matter of weeks before he would be beating the best. This is a nice dream, but if the budding champion ever has a chance to ride top-class machinery he discovers that these champion blokes have an awful lot more than good bikes on their side: riding ability, super-human strength and fearless courage, to name but three helpful traits. As a professional observer of the MotoCross scene, I was all too aware of these qualities which make a star, but even so, I had this sneaking subliminal idea that I too, had the makings of a champion. After a couple of hours aboard Malcolm Davis' latest works machine, I can say with some assurance that I am never, ever, going to be a challenge to the established Grand Prix clique.

These days there are two types of professional rider employed in MotoCross. One is the man who is there almost exclusively to win. His lot will be to race the most competitive machines a factory has available and to earn his keep by winning as often as is superhumanly possible. He is always employed by the bigger and therefore richer factories, who can afford to pay him. In the second bracket it's the rider who has to be talented enough to both win at a very high level of competition and at the same time develop new machines for the factory. He is both a works rider in that he is expected to win, and development engineer because he is expected to help with the production of new models at the same time as he does his winning. His is a hard life.

So it was that Malcolm Davis found himself presented with a production prototype 400 Bultaco which he is to develop and refine during 1973 for production in time for the 1974 season. It is

TRACK
VILAR

important to remember that the bike I rode was a PROTOTYPE in a relatively early stage of development and it was only through the good offices of Malcolm and the Bultaco factory that MCI became the first British magazine to try the bike.

Rumours of Bultaco's 400 first began drifting out of the Spanish factory about 18 months ago. At the time, Bultaco enjoyed a reputation for building some of the fastest 250 scramblers in the world—even if the machines didn't bathe in the same rosy glow regarding bottom-end power. Then there was the ferociously fast 360cc Bandito; dogged by atrocious handling and gross over-weight, but still an absolute flying machine in a straight line. So what could a factory with this reputation do with a full 400cc? The MotoCross world watched and waited.

Soon stories were heard of frames being twisted inside out and gearbox internals being shredded by the new power-plant. Test riders at the factory gave up trying to tame the beast and the designers went back to their drawing boards.

Development continued and one or two "hack" motors were released for riders to assess. These engines, being very early prototypes, were not very reliable and more re-thinking was done. The culmination of all this effort was the prototype, and once again I would stress that the bike is only a prototype, 400 we tested for a couple of memorable hours in the rolling Gloucestershire countryside.

I had been looking forward to riding the big Bul ever since Malcolm had mentioned that it might be possible for me to ride the bike, but when the time arrived I wished I was somewhere a long way away and some other rider, braver and with more ability than I, was going to have the privilege of being the first British journalist to ride Bultaco's latest challenger.

The motor, unpolished and rough, looked very menacing as it crouched low down in the frame. Its peeling, black heat dispersant paint created the impression of a Royal Marine ready to go out on night patrol and the unfinished castings seemed to add a menacing air of unsophisticated brutality. The Bul had me really worried and I hadn't even got the thing running yet.

Malcolm did this for me by swinging down on the forward-mounted kickstart and the 400 crackled into life. In an unsilenced state, Bultacos are always noisy, but this one was positively deafening. I would guarantee that there isn't a noise meter made that would even record how much racket this machine makes, it is too much by far. Needless to say, Bultaco are working on this problem as well. (Ed note: Bultaco have asked us to point out that the 400 is now silenced, as are all their British works bikes.)

When I summoned sufficient courage to actually sit on the bike, I was a little more re-assured. It felt much the same as the Pursang I rode last year which is hardly surprising since the cycle parts are almost identical to the present production 250. This meant that everything fitted me nicely; or in other words, I could reach both brakes comfortably. And at this stage of the game, brakes were what I was most interested in.

Malcolm promised me that the power was not uncontrollable, although he did stress that there were an awful lot of big, strong horses riding inside those rough cast crankcases. Caution, he advised was necessary, and I am not one to shun advice from a rider of Davis' calibre.

I could delay no longer. One can hardly bring the whole British Bultaco team and a photographer together merely to stand watching a quaking wretch blipping the throttle of the bike which he felt sure was going to kill him. So, in with the clutch, a few revs and off we went. Not as I had thought, in a death defying arm-wrenching lurch, but quite sedately by competition standards.

About three minutes later I had discovered that the Bultaco was not at all a savage beast; rather an obedient but immensely powerful creature. If you want to drive the big Bul around at 600 revs in first gear, the bike is quite happy to do this. Should you want to go a little bit faster, you open the throttle just a fraction. But give a big pull at the right-hand grip and the Bul will simply and unashamedly take off in an explosion of wheelies, slides, drifts and anything else that can conceivably happen to a MotoCross machine.

There is no use waiting for a time lag or even the most miniscule breathing space between cracking the throttle open and the bike responding. You just HAVE to be ready and in control, because once the 400 is given its head, things happen awfully quickly.

I found this out after only two laps of our test track, when I sat too far back on the saddle for a very simple jump, and landed with just a shade too much throttle. On a normal top-

OVER

THE 400 BULTACO'S TRACK TEST

class machine,
this would have meant that the front wheel would have been a trifle too high on landing, but certainly nothing more. The Bul took off in a heart-stopping, near vertical wheelie, which took me about 30 yards to sort out. After that, I treated the bike with an extra special respect on jumps.

It would have been to my advantage to have extended this ultra-respect to other parts of the course, but a little too much enthusiasm led to the supreme embarrassment of dropping a "works" bike. The Grass-Tracker in me likes fast, sweeping bends and I had really settled down to drifting the big Bul through a quick, but bumpy right-hander at the bottom of the course. After a couple of hours riding, I reached the state where I could press the bike fairly quickly through this corner, in a controlled slide. At least, it was intended that the slide was controlled. Clipping the corner of a large, deeply embedded rock and the tiniest amount of excess throttle and P. V. was sitting on his bottom looking at a recumbent works Bultaco. (Ed's note: Serves him right for taking off the training wheels.)

The incident served to illustrate the bike's immense potential and also the essence of its character. There is nowhere on any MotoCross circuit I have ever seen that the big Bultaco would not unhesitatingly accelerate, and accelerate with extreme rapidity. But, and this is a bit but, to get the Bul to perform at anywhere near its limit, is going to take a rider of the calibre of Malcolm Davis.

Unlike some works machines that I have ridden, the 400 was not difficult to ride. For example, the power didn't come in with a nerve shattering surge at any particular point in the rev band. On the contrary, as I said earlier, if you want to go slow you keep the throttle closed, and the faster you want to travel the more you open the throttle. It is as simple as that. But racing the thing is going to be another matter.

It is all very well testing a bike as I did, with no-one else on the track to disturb my concentration, and no reason to press my riding ability to the farthest edge, but in the heat of battle, things would be quite different.

Malcolm was the first to admit that the bike is almost a new concept in riding, simply because even the works riders have not had the sort of power available that the 400 offers. One has to re-think every racing line and acceleration point to make allowances and — if the rider has the ability — make use of the immense reserves of power.

The motor which produces this gentle, but vast amount of urge, is quite different from any Bultaco either past or present. To begin with, it is by far the biggest engine the Spanish factory has ever produced and it is also quite a departure from the Pursang / Sherpa / Matador / Alpina range which has won so much acclaim.

The most obvious departure from convention is the forward-mounted kick-start — something of a rarity in the motorcycle world, but a design proven in the world's most reliable cross-country bikes — the ISDT Jawas. You may rest assured, if forward-mounted kick-starts appear in a Czech Trophy team, then they are a GUARANTEED success. The main advantage with this mode of starting the engine is that when the pedal is not in use, it points forward, folding well out of the rider's way.

In practice, the system worked well. One pulled in the decompressor lever and took a hefty swing with the right boot. Once the engine was nicely spinning — approximately half way through the pedal's travel — the decompressor was dis-engaged and the engine usually fired. The kick-start then rested well out of harm's way in a forward position.

Although no specific details were available regarding the engine — even Bultaco aren't going to open up their secret engines for MCI — Malcolm was able to give us some superficial information.

The engine is under 400cc and at the time of writing gives 42 bhp at 6,000 revs. The early motors gave as much as 48 bhp but even works' riders can have too much of a good thing.

Although this is pure guesswork, I think that the engine is a long-stroke design, because the cylinder-head is very similar in size to my 250cc Pursang, and if the 400 were a really big-bore configuration, it would appear that a noticeably bigger casting would be needed for the head. But as I said, the last statement is pure speculation.

Without guessing, I can tell you that the 400 inhales through a 38mm Amal Concentric carb. At present a 480 main jet is used and the engine still runs weak, so the breathing must be exceptionaly good.

The finning follows current Bultaco fashion in that it is squarish in appearance with anti-distortion bays machined in at staggered intervals. Again, very much after the Pursang design — only more fins.

In the head live two Champion N3G plugs which are fired simultaneously by a Femastronic electronic ignition system, the reliability of which will be testified by thousands of Bultaco owners.

Moving to the bottom-half of the unit, one finds lots of new and exciting goodies. To begin with, we have the striking starting mechanism I mentioned earlier, but in the engine cases there also lives a gear primary drive and multi-plate clutch which uses bonded clutch plates — similar to those found on British super-bikes — instead of the more usual metal items.

Blast off! Britain's answer to Skylab. Malcolm Davis shows how it's done.

The clutch action is worthy of mention because it was so incredibly light — especially when one considers the amount of power it is handling. When I rode the bike it neither slipped nor dragged, but there were dark murmurings in the team about "work" having to be done on some aspects of the design. It is this kind of perfectionist thinking which pays dividends for the private owner.

The gearbox was less pleasing. Changing from first to second was excellent, regardless of whether one was coming up or down the box, but there was a long, long movement to engage third gear, and this took some getting used to. The new gearbox has only four ratios, but this is more than ample. I was too cowardly to use top, and bottom would produce enough torque to pull down Honda's sales graph, so for the vast number of riders the big Bul will be a two-speed racer. These two gears, for the doubters, will take you from a standing-start to 50 mph in less time than it takes to say Bultaco. And 99 per cent of all MotoCross racing is done in this speed range.

I have deliberately avoided mentioning the bike's cycle parts because the paint was hardly dry on them when I rode the bike, and there was obviously a lot of development work to be done. In a case such as this it is merely fatuous to proffer criticism since things are being changed by the minute. The chassis, with the exception of the engine mountings, is pure Pursang and hence it is hardly surprising that both Malcolm and I found the suspension far too soft. But such things are mere trivia, and in view of the early stage of development that the bike was in, it handled extremely well. Much after the

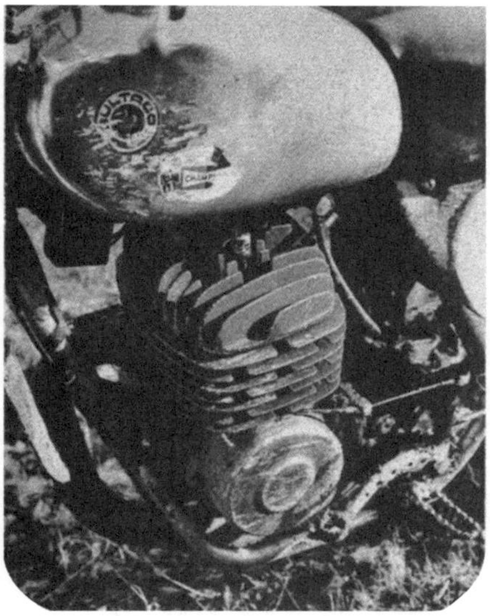

Bul's head is fitted with decompressor and twin plugs which fire together.

Overworked Betor forks were satisfactory, aluminium alloy hub is incredibly light.

If you want to ride round sedately, the Bul is happy to do so.

fashion of the super-powered 250 — which is what it is at present.

There were two items of interest which are worthy of mention since they will be fitted to production models in 1973. First was the new front wheel which featured both an alloy brake part and a tapered alloy hub. The brake was still a good one, and the. latest slimming exercise must make it one of the lightest in existence.

Also new were the forged alloy fork yokes which are designed to eliminate the oft criticised twisting of the fork legs, which Betor forks are rather prone to. The new yokes look nice as well and should give the polishing enthusiast another little challenge. I don't know whether they will be any better, but the idea seems sound, and they look pretty so one can't really ask for more.

To come to any firm conclusion about the 400 Bul is rather difficult because it lacks comparison. It is, to the best of my knowledge, the fastest MotoCross machine I have ever ridden. I say "to the best of my knowledge" because memories fade with time and once concrete beliefs begin to distort. Even so, I am fairly sure that of all top-class machines I have straddled, none was faster than this bike.

More than being the fastest machine on the track, the 400 has great sales potential in that it can be ridden by an average rider, which is not the case with a number of current works machines.

The forward-action kickstart can be seen clearly in this shot.

Whether you want to risk your neck on a bike which has such potency is another matter, but as Malcolm says, "It feels good when you know that you've got the quickest bike in the race and all you have to do is ride it." It is this attitude which will sell the 400 Bultacos, for it is going to be a real M-A-N who is going to find this bike too slow for him.

P.S. Regarding the Bul — my name is Petula Vilar.

BULTACO – the bike for all seasons

125 BULTACO PURSANG

ANOTHER SCREAMING SPANISH ROCKET

Amal carburetor comes with a tickler instead of a choke. It worked very well.

Betor forks are beautiful in both appearance and performance. Polished Akront rims are standard. Fiberglass fender should be plastic.

The Bultaco is a maze of red, grey, and chrome parts. It is an attractive motorcycle.

In 1972 Bultaco dropped the Sherpa S line and renamed all of their smaller short track and motocross bikes Pursang, like the larger models. The reason why is only known in Spain, but the name Sherpa never really fit any of the competition machines. Sherpa means "mountain goat" and could only be effectively applied to their World Champion trials bike. The new designation wasn't the only thing that Bultaco changed last year. Along with it came a complete facelift and internal rebuild. The Bultaco Pursang is the acknowledged 125 class speedster. This year the machine comes with redesigned porting, a smaller carburetor to tame down its wicked powerband, and a new set of triple clamps which keep the front end in place eliminating the famous Bultaco wash out.

The Pursang's engine is unique. It is a 250 mill with a smaller piston, cylinder, head, and carburetor. The exhaust pipe also differs. Because of this arrangement, the smallish 51.5mm piston travels 60mm each stroke. Small bore and long stroke is not a combination that would be termed conducive to high rpm. Nevertheless the Bul claims its maximum horsepower of 25.1 all the way out at ten grand. And it'll rev higher if you push it. At ten grand the piston is traveling at well over 3900 feet per minute!

Primary drive is by chain rather than the conventional gear method. The only other off-road motorcycles that we can think of which use primary chains are the Maicos. Power delivery is smoother with a chain. But they do wear out and snap, especially at the excruciating rpm the Bul likes to run at. With the increased horsepower over previous 125 Bultacos, the primary chains became a problem in '72. They were snapping left and right often lunching the side case as well. Late that same year Bultacos began arriving with wider gears on the primary hubs and a stronger chain to match. This was carried over to the '73 model we tested and it works just fine. The chain does have to be replaced periodically. But they don't wear out as rapidly as they did. The 27mm Amal carburetor this year is three millimeters smaller than last year. Power is still surgy, but in a much more controllable manner. Starting is also facilitated and the bike won't load up as easily. The right side shifting gearbox works like it was built for a steamshovel. Crisp shifting in either direction and strong as an ox. This is the last year that Buls will have a right side shift as Federal legislation will make left side shifters mandatory on all machines sold in the States in 1974. Ignition by Femsa is an electrical pointless unit. With the spark that is required at ten grand, it would be difficult to keep a conventional set of points properly adjusted for very long.

The frame is manufactured from what other factories like to refer to as "water pipe." Mild steel. It is strong. Not because mild steel is exceptionally strong, but because Bultaco uses thick tubing and well placed gussets in the frame. Much of this is responsible for the overall weight of 217 lb. A single downtube splits beneath the exhaust port to form a cradle for the engine. The long swinging arm is mounted to a plate just inside the engine's quarters. The whole chassis is painted in hammertone grey.

Forks are manufactured by Betor, who have been making Bultaco's forks since way back when. They work as well as a

Long stroke mill really likes to put out the Rs. Serrated footpegs are spring loaded and self cleaning. Shifter will be on the left side next year.

Skyway spark arrestor/muffler keeps things quiet. Betor shocks have very poor downstroke dampening and bottom easily. Chain adjustment is by notched snail cam.

fork can work, although they did ooze some oil out of the top filler nut. Betors are the favorite forks of Joel Robert; and, according to Jim Pomeroy, Joel said that if the Japanese could build forks like the ones on Bultacos, he wouldn't be having any handling problems with his Suzuki this year.

There are two sore spots (one of them is literally a sore spot) on the little Pursang. The shock absorbers and the seat. Betor also manufactures the shocks. Last year they worked beautifully, if you could keep seals in them. This year they solved their leakage problem but they have developed another problem which is even worse. Rather than utilizing better seals to keep the shocks from blowing up, they redesigned the valving inside for longevity. The result is shock absorbers that can be bottomed out by a 100 lb. rider. Bad choice, Bultaco. The seat is the work of the Devil. In order to fit the seat over the rear fender, Bultaco contoured the seat base to the fender. They did the same thing on last year's model only this year the seat base is further away from the fender. What was created is a hard lump in the seat exactly where you sit when sliding back for more traction. If you aren't thinking, it can give you a solid whack in the rear that will soon have you at your dealer's, purchasing a '72 seat.

Wheels are Bultaco's own. The rear hub is a heavy cast item. Conical in shape, it contains a fine brake. The conical hub up front is dynamite. The stopper is a two finger grabber. The hub itself is very light and will be even lighter when Bultaco begins using the hub from their Sherpa T. It is an eggshell thin binder manufactured of Dur-al. Both wheels carry Akront rims. The Akronts may pack mud in, but the mudless alloys aren't available in Spain yet. When they are, you can be sure that Buls will have them.

Externally, the chrome pickle is gone. That super effective two-ton muffler has been replaced by a Skyway unit. The Skyway is almost as effective, weighs much less, and looks a whole lot better. The fuel tank and side panels are the same as last year with their red, grey, and white styling. The fiberglass fenders, which are good for about three Sundays before they vibrate themselves off, are still with us. They work exceptionally well at keeping mud and other debris off the rider and engine, but they are brittle and won't flex as much as popular replacement fenders. The handlebars are exceptionally wide—35 in., but they are designed to be cut down to rider preference. Alloy levers are standard as are Pirelli tires.

Previously, when a Pursang hit its powerband, all sorts of things happened. The rear tire spewed dirt; the front end danced lightly, and the bike had to be near vertical or the rider must be prepared for a wild slide or even a spill. Now you can feel the power coming on. It doesn't slam you in the back, but is smooth and predictable. A welcome change. The other change that we liked is the front end geometry. New polished aluminum triple clamps alter the Bultaco's rake and suck in the wheelbase. Last year it sported a 55.9 in. stretch. Now it is only 54.5. Fear not, that inbred Bultaco steadiness is still there, but now you can corner without having to sit on the handlebars. You do have to slide your weight forward, but it is a beautiful change from previous Buls. Remember the ones that cornered like a Kenworth on ice?

Power is more than ample and once the starting gate drops, you can forget about the clutch. If you are used to other machinery, it will take you some time to coordinate the right side shift with a left brake. But it can be done. On extremely dry or muddy surfaces you may wish you had other tires, but for a tacky track the Pirellis work fine.

For the price you shouldn't have to spring for better shocks or last year's seat. But if that's all it takes to make your Bul ready to win, be content. Other so-called motocrossers need a whole lot more.

Left side of the rear hub is a busy place with the brake and rod, sprocket and chain, shock, chain guide/anchor arm, and adjuster.

Light conical front brake is a real stopper. The spokes are reinforced at the hub flange.

BULTACO 125

DYNAMOMETER TEST REPORT

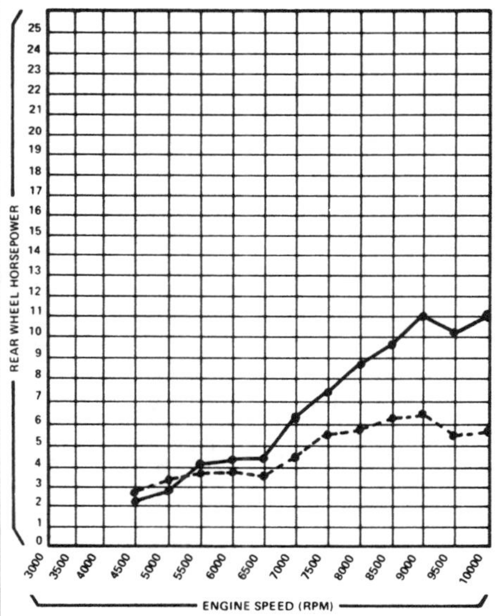

RPM	TORQUE
10000	5.7
9500	5.5
9000	6.4
8500	6.1
8000	5.7
7500	5.3
7000	4.6
6500	3.6
6000	3.9
5500	3.8
5000	3.1
4500	2.8
4000	
3500	
3000	
2500	
2000	

TORQUE - - - - -
HORSEPOWER ———

SPECIFICATIONS

PRICE

Suggested retail $995

ENGINE

Engine type . . . . . . two-stroke, piston-port, single-cyl.
Horsepower @ rpm . . . . . . 25.1 @ 10,000
Torque @ rpm . . . . N/A
Bore and stroke
in. . . . . 2.02 x 2.36
mm . . . . 51.5 x 60.0
Displacement
cu. in. 7.66
cc 124.98
Compression ratio 14:1
Carburetion . . . . . 27mm Amal
Ignition . Femsatronic pointless

TRANSMISSION

Speeds . . . . five
Primary drive . . . duplex chain
Clutch type . . . . wet, multi-disc
Final drive . . . chain, 5/8 x 5/16

CHASSIS

Length, overall, in. 82.6
Wheelbase, in. 54.5
Ground clearance, in. 9.6
Weight, overall, lb. 217
Frame type . . . . . . single cradle
Tire size
front 3.00–21
rear 4.00–18

A proven winner for the TT specialist.

It shouldn't be necessary to introduce the name Bultaco to anyone who has followed motorcycling for any length of time. Bultaco made a name for itself back in the good old days when sportsman racing was the largest challenge to the non-professional Sunday racer. At one time, scrambles were as predominant as motocross is today, in fact more so. And at every scrambles course you could almost count on seeing a Bultaco of one size or another in the winner's circle.

Aside from Bultaco's sportsman racing effort, they also build several different-sized machines that are competitive for the increasingly popular enduro events. Instead of introducing a bevy of entirely new motorcycles each year, Bultaco seems to concentrate on improving their already competition-successful models. The Spanish concern is a relatively small company; mass production is not their game. They seem to have no trouble selling the quantities their facilities are capable of generating.

Since Bultaco is quite active in international competition, ranging from trials (ever hear of Sammy Miller?) to motocross (Senor Jim Pomeroy) to the International Six Day Trial, they receive sufficient feedback to refine their production units. They don't really need to revamp them, they've been at the game long enough to have the basics pretty well shaped.

This year's 200cc Pursang Mk 6 is quite similar to earlier Sherpas. The only noteworthy change, besides that of the name, is in the powerplant. A Matador (Six Days Trial) top end has replaced the standard Sherpa unit. Appearance wise, the Mark 6 seems no different. The color arrangement remains Bultaco's ever familiar red and gray.

Fiberglass is used for both front and rear fenders and the two gallon fuel tank, instead of the more popular and durable plastics. Fiberglass has a tendency to de-laminate and is more prone to breakage than the resistant plastics which can be twisted, bent, and banged, and yet still return to their original shape. We feel that this unbreakable plastic is the way to go, and hopefully someday soon all off-road machinery will be so equipped.

The single cylinder two-stroke en-

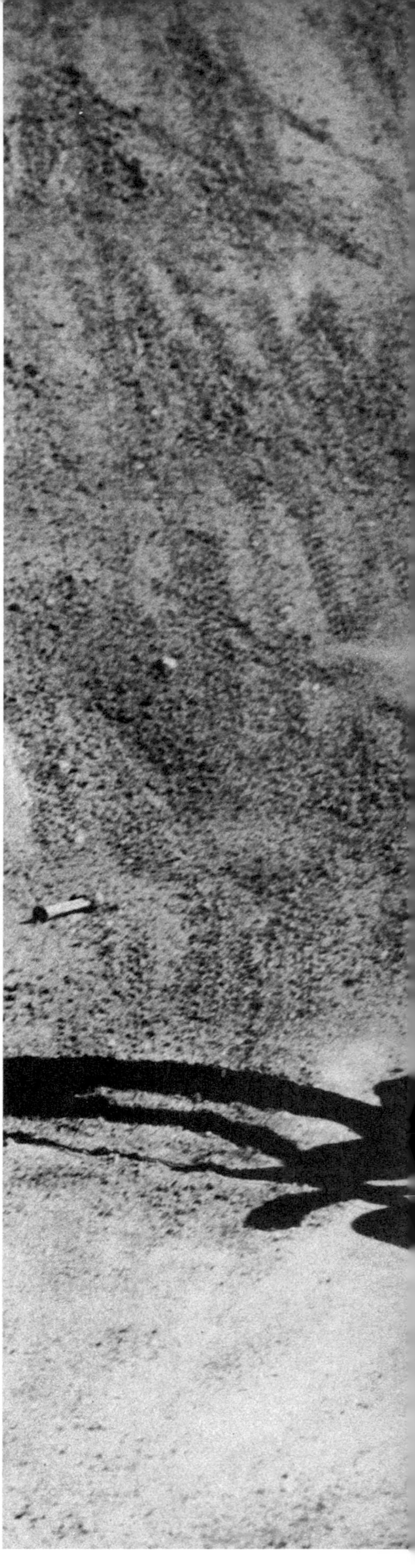

gine does not have oil injection. Gasoline and oil are mixed at the ratio of 32 to 1. With a bore and stroke of 64.5 by 60mms, total engine capacity is 196ccs. The uncorrected compression ratio is 13.0 to 1. Bultaco does not use a reed valve. They utilize a very conventional piston-controlled five-port cylinder design. Bultaco claims the engine puts out 31.2 horsepower at 10,000 rpm. We feel that this figure is extremely optimistic. The 200 has some ponies, but not that many.

Perhaps the most difficult thing for some riders to adjust to is the gear shift lever positioned on the right side of the engine. The five-speed constant mesh transmission was surprisingly smooth and positive. This, in part, can be attributed to the short travel of the shift lever. Unlike some Bultacos we have ridden, we were not plagued by any false neutrals or over shifts. Once the gear change was made, there was never a second thought about that gear becoming disengaged. Even after a hard go around on the racetrack neutral was still very easy to find.

The clutch is the wet multi-plate variety. It is a carry-over from Bultaco's water-cooled road racers. Unlike 99% of the clutches found today, the Bul's clutch utilizes steel plates alone. There is no friction material bonded to either the driving or driven plate. The clutch required an average

CYCLE
RIDER

Strangely, the handlebars seem more like motocross bars than TT bars. They seem a bit high and forward.

The handgrips were comfortable and did not force us into using a death grip on the bars. The lack of vibration is another plus point in favor of the Bultaco.

Unlike the knobby or trials universal tires fitted to most dirt bikes today, the Bultaco uses a Spanish-made Pirelli tire with a block type design. This, in itself, is a dead giveaway as to the purpose for which this machine was intended. This type of tire is designed for a surface that has a consistency somewhere between a hard slippery groove and a deep loose cushion. The tread can also be cut by razor blades and adapted to various track conditions, or to compensate for normal tire wear. Up front is a large 3.50 x 19 tire on an Akront aluminum rim. A 4.00 x 18 is found on the rear.

Bultaco, and Spain in general, has a reputation for less than inspiring brakes. The 200 does nothing to alter that reputation. The front single leading shoe stopper does not require an excessive amount of pressure before it starts to take hold. The performance, however, is nothing spectacular. We were actually unable to skid the wheel by pulling the lever clear back to the handlebars. We did notice, though, that when braking hard, the front end did have a tendency to nosedive. The rear brake was more of the same, but worse. It required an excessive amount of pressure, like standing on the lever, before the rear wheel would lock.

It was just short of impossible to use this brake hard when negotiating extremely rough sections of track. Aside from the dismal rear binder, braking was also hindered by the fact that the rear end would leave the ground. A floating rear brake, along with better suspension, would improve the situation. Admittedly, we were using the Mk 6 on a motocross course that was somewhat rougher than the majority of TT scrambles courses throughout the country. On a second, smoother course, the brakes, both front and rear, performed much better, but they were still nothing to get excited about.

Aside from the slight problem we encountered with the front end bottoming out under hard braking, we were quite pleased with the performance of the front forks. Our unit was tested on two courses of varying roughness. Even so, the front suspension smoothed even the roughest of sections. We can't get as enthusiastic about the rear suspension. On a smooth, fast course the rear shocks were satisfactory, but on the rougher course the shocks lacked the damping control that we feel is necessary for top line handling.

The shocks are five-way adjustable. We found it necessary to set the preload up to the fourth position. This prevented the shocks from bottoming out, but because of the preload the damping was lacking.

The power performance of the engine is suited for one particular type of track, one that is smooth and rather fast. The 200cc Pursang did not have an over abundance of low end performance, but the power came on strong in the mid-range and continued right up to what we imagine to be maximum rpm. From that point on there was a gradual fall off in power. Once above the low speed sluggishness, the throttle response was quite good. The power does come in smooth, and doesn't exhibit a tendency to spin the machine out when accelerating hard out of a corner. On a smoother course there is no question that the Mk 6 handles quite well. We feel that it is possible to maneuver the machine in any manner that is suited to the rider. In other words, the power characteristics of the engine are such that the 200 can be slid through a corner with the throttle on, or it can be tracked around with equally good results if the corner is hard and slippery. This performance is a direct function of good frame geometry.

In conclusion, the one thing we can say about the 200 is that it is a proven winner. Bultaco's have dominated this class size in a manner much like Yamaha dominates the 250 class in professional road racing. When looking at the bike objectively, there is no question that it will accomplish the job that it was produced to do with very little effort. It could be refined even more. Maybe one of these days Spain will come up with some good brakes, and Bultaco will market a comfortable racing saddle. While handlebars are easily changed, those on the stocker seem motocrossish to us, better suited to standup riding. After innumerable tests calling for serrated footpegs on the popular dual purpose bikes it seems peculiar to be voicing a preference for rubber covered pegs on the 200. But AMA sportsman rules do require them for TT-type events. Right now its just another item that has to be changed by the purchaser.

Aside from these complaints, it is easy for us to see why the 200 has achieved the success it has. The bike is really good for only one thing, but if TT is your thing, then you're looking at a very good bike. ■

BULTACO 200 PURSANG MK 6

Engine type	single cylinder two-stroke
Bore and stroke	64.5 x 60mm
Displacement	196.04cc
Compression ratio	13:1
Claimed horsepower	31.5 hp @ 9500 rpm
Engine red lines @	9500 rpm
Ignition	CDI
Starting system	kick, in neutral only
Carburetion	32mm Amal
Lubrication	pre-mix
Type of transmission	constant mesh 5-speed
Clutch	wet multi-plate
Overall gear ratios	(1) 24.14, (2) 17.73, (3) 13.58, (4) 10.91, (5) 9.10
Final ratio	9.10
Countershaft sprocket	12
Rear wheel sprocket	46
Length	82.6 in.
Seat height	32 in.
Wheelbase	55.9 in.
Ground clearance	8.25 in.
Listed dry weight	216 lbs.
Actual weight, full tank of gas	238 lbs.
Front tire size	3.50 x 19 in.
Front brake type	internal expanding
Rear tire size	4.00 x 18 in.
Rear brake type	internal expanding
Air filtration	washable foam
Fuel tank capacity	2.0 gal.
Gear box capacity	600cc
Front suspension	telescopic double damping
Rear suspension	5-way adjustable spring over shock
Frame type	semi double cradle
Exhaust system	silenced expansion chamber
Color	red with silver trim
Retail price, Los Angeles	$1040

DISTRIBUTOR
Bultaco International Ltd.
P.O. Box 62452
Virginia Beach, Virginia 23462

It has parts from the motocrosser and handling like the trialer.

BULTACO'S MATADOR MK V

Equipped with everything needed for an enduro even more power from last year's Matador.

If you were to categorize this machine as a cowtrailer, desert sled or plunker, you might have to start a new category—maybe like mountain-goat trailer. In this day and age of specialized motorcycles, where you seem to need a special machine for each type of competition there is this entry to the *enduro only* bracket from Bultaco. It's Matador MK V is a 250cc single cylinder two stroke enduro machine that will surprise you with what it can do.

In the Matador's career, which stems from 1962, the Bultaco has netted over 35 Gold medals in International Six Day Competition. Obviously these trials and results are why the Matador is current and above the competition.

It's not really a street and trail machine as one might relate to the

Yamaha Enduro or similar bike. This machine doesn't have a lot of fancy trimming like mirrors, turn signals or battery for the headlights. It can be ridden on the streets with the addition of some of this equipment, but then it would be too heavy to do the job it was intended for, and that's to cover any trail possible that one might encounter during an enduro. In its gassed and ready to go trim the machine weighs 260 pounds.

The Matador MK V reeks with features from all the Bultacos: engine parts from the motocrossers, steering near the trialer and refinements from past Matadors and Alpinas.

The new features, starting from the front, are new shoulderless Akront rims. Akront rims in the past have proven to be one of the strongest available for their weight but they had the raised shoulders which were famous for collecting mud. In most enduros you could count on mud and that just made the rider's job more fatiguing. The DID rims were being used by most people who were in the mud a lot but now with the new shoulderless Akronts you have the best rim and that means one less part you have to spend extra bucks on once you have made your purchase. The rims also have locking pins or screws in the side of the rim to hold the tire from spinning. This replaces the rim locks and saves weight and an unbalanced tire. When you change the tire you do not remove these screws.

The rim is laced to a Dural alloy front hub which is lighter than last year's. It was taken from the motocross machines with the addition of a speedometer drive unit. The front brake lever has a booster spring on it to assist the release of the brakes should the front end become heavily packed in mud. Normally when this happens and you apply the brakes they either don't release or else they drag.

New triple clamps are from the Pursang MK VI. Their use is to give the machine more trail. They also move the handlebars further forward which will be noticed as more comfortable than previous models. The forks themselves are the same as '73, Betors with 6.5 inches of travel.

New cables have been used; these with a nylon lined cable housing. The levers all have mud guards to protect them from mud packs and the throttle is well protected, too. The throttle also features a new guide to prevent the cable from bending where it exits the twist grip.

The speedometer is a VDO that is adjustable to a tenth of a mile. There is also a large magnifying glass and watch for easy time keeping along a route. A route reel is also standard on the Matador.

Moving aft we find that a new 2.8 gallon fiberglass gas tank is used with a super sealing plug type gas cap. The cap is for quick removal during a pit stop but then one might not appreciate the small opening to pour the gas. With the large GI type gas cans the fit is tight and doesn't allow you to see how filling the tank is going. The cap has a breather hole in the top and it too leaks gas when the tank is full. Under the tank are two petcocks which is a new feature. The new Matador engine required the new arrangement as the 32mm Amal carb

Knobby tires are standard as is the latest trick from the Akront wheel people, shoulderless aluminum wheels. No more heavy mud packing!

The new gas cap is supposed to be leakproof. It worked well on the trail, but when the tank was overfull it spilled some gas through the breather hole.

had to provide better fuel flow. The petcock on the left side of the machine is a *bear* to get to as the exhaust pipe leaves little room between the tank and pipe in which to get your hand.

Now for the engine department the Matador MK V really shows some improvement. The top end is from the Pursang MK V motocrosser. Some might think that a motocrosser might make a poor enduro machine but in this case the combination works well. The engine is not pipey nor does it have such a powerful top end that it will win you any motocross races. Somehow Bultaco has managed to balance the low-end and top-end power so you have a good broad powerband that works hard constantly.

The lower end of the engine has remained unchanged with its strong characteristics of a ball bearing support crankshaft and a dual row primary chain.

The carburetor gets its air supply from a sponge type filter which is inside a well shrouded airbox. You'll have to sink the Matador engine deep into the water before drowning it.

Right behind the airbox is mounted an air bottle with tire sealant. It's obvious what that is for and unfortunately during one of our test rides we had the chance to test the setup. The only problem was that we put a rock through the sidewall of the tire and the sealant didn't offer any help. We have seen this sealant work well with smaller punctures.

The rear hub is another new item for the '74 Matador. It is four pounds lighter than the unit last year. The swing arm has a cam type adjuster for the chain which helps assure accurate alignment of the chain and wheel. It is hard to sight the chain for alignment as it is fully enclosed which eliminates the possibility of mud and rocks from breaking the chain. The shrouding looks bulky but it's worth the weight which is really not noticeable.

When climbing aboard the Matador you will notice the new seat padding. The front of the seat at the tank area is raised slightly. The previous seats did not have this and riders complained that it allowed you to slip up against the tank, placing too much weight on the front end.

Starting the Matador MK V requires you to tickle the Amal carburetor until it is overflowing. Crack the throttle slightly and then pull in the clutch lever to bring the kickstarter to the back of the engine so you can kick it through. The engine never started on the first kick when it was cold but it was never a problem to get it running within three or four kicks. When it was warm, first kick starting was every time. Usually test riders have commented that bikes equipped with Amal carbs should be changed to something like a Mikuni for easier starts and better performance. With the Matador, the Amal does a fantastic job allowing the engine to pull very strong from low-end on. It would be hard to see any benefit by changing to any other carb.

The engine features a motocross top end but power is very smooth and controllable.

We tried the flat repair stuff but we put a gigantic hole in the tire and tube so the air didn't hold up long enough to get us the ten miles back to the truck.

The rear hub is four pounds lighter than last years. The bottle is filled with sealant and air for flats.

The kickstarter is in an awkward position for riders other than those who are Bultaco freaks. But the location does put it completely out of the way while riding. Another problem for a non-Bultaco rider is that the shifter and brake are on opposite sides from standard. This will be changed in 1975 as that is when the Federal regulations requiring the standardization of controls goes into effect.

During our riding we found that the Matador takes some time to get used to and feel comfortable. You tend to sit with your legs further back than normal and the tendency is to stand up to feel comfortable. Actually the problem we were having was to sit in the normal pocket of the seat which is why we were uncomfortable. Once we learned to sit further back, to the point where you felt the raised padding on the back of the seat, everything was fitting better. When you do stand up while riding you feel high. The pegs are 9.6 inches above the ground which gives you good clearance from rocks or other high hazards.

When you're riding a Matador you'll love the torque and power for trail riding. It will haul you around the mountains and up the hills with a very maneuverable powerband. It turns like a trialer in a sense meaning the steering is on the quick side. For high speed rough terrain like a desert race, the Matador won't handle as well as a bike more suited for that type of racing. The suspension is good for the trails but again it won't be beneficial to the aforementioned high speed stuff.

We found the front end gave you the feeling that it was bottoming very easy. Checking the travel of the forks by the marks on the legs we could see that we never did bottom them but the tire was merely hitting the fender making a loud scraping noise. The fenders are new and made of unbreakable plastic. The brakes are something that we had inconsistent feelings about. Sometimes they worked great, other times not so great. It seemed that the front brake worked better at high speeds but the back never would lock up.

The Matador MK V has some problems but none with the basic design or performance. It is a super machine in the hills and a perfectly prepared enduro machine requiring nothing more than filling the tank with gas to be competitive out of the crate.

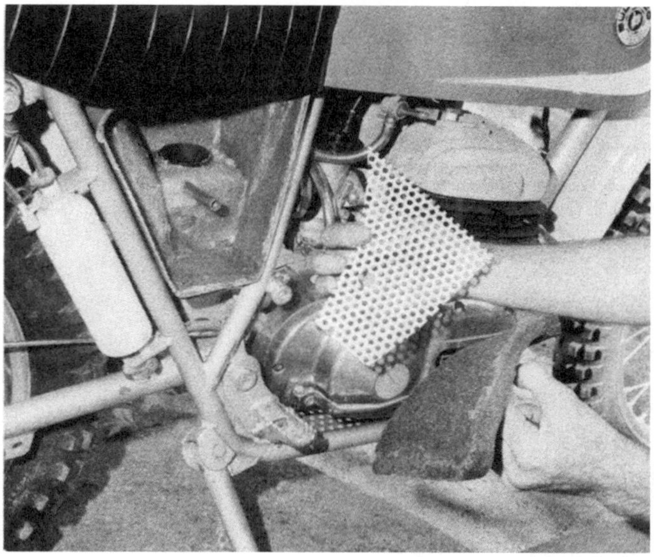

The air cleaner is very efficient and protected from deep water crossings.

MATADOR MKV

Suggested Retail Price: $1195

ENGINE

Engine type	two-stroke, single cyl.
Bore and stroke, mm	72mm x 60mm
Displacement, cc	244
Horsepower/rpm	29.5/7000 rpm
Compression ratio	10:1
Air filtration	air box/foam element
Carburetion	32mm Amal
Lubrication	gas mix
Ignition	flywheel magneto

DRIVE TRAIN

Transmission	five-speed
Clutch type	wet
Primary drive	dual row chain

CHASSIS

Chassis type	double-downtube
Seat height, in.	33
Peg height, in.	13
Handlebar width, in.	34.25
Ground clearance, in.	10.5
Wheelbase, in.	55
Weight (as tested), lbs.	258
Tires, front	3.00x21
rear	4.00x18

Max. Pts.	NUMERICAL EVALUATION	
10	Starting	8
10	Power	8
10	Powerband	10
10	Transmission	
	(5) Ratios	5
	(5) Operation	3
10	Suspension	
	(5) Front	4
	(5) Rear	4
10	Brakes	
	(5) Effectiveness	3
	(5) Waterproofing	3
10	Handling	
	(5) Low-speed Maneuverability	5
	(5) High-speed Stability	3
10	Comfort	8
20	Miscellanea	
	(5) Quality of craftsmanship	5
	(5) Instrumentation	5
	(5) Electrics	5
	(5) Noise level	4
100 pts.	Overall Rating	pts. 83

Feature Editor Brad Zimmerman demonstrates the common way to bring a Bul to life. Stand on the left, tickle and Amal and give it two swift kicks. The machine started easily after we removed the optional in-line fuel filter.

It looks plain, but racy. Tipping the scales at a mere 204 lbs., the new Bultaco Mk VII 360 is definitely a serious out-of-the-crate competitor, able to win races in relatively experienced hands.

BULTACO'S PURSANG Mk.7

Ever since Jim Pomeroy won the Spanish Grand Prix last year, people have been looking up and taking note of Bultaco motorcycles. Of course, they've been around for some time now, but many riders didn't even consider the bikes for competition. Bultaco is a small company, thus doesn't have the backing to go to inordinate measures to advertise their machines. But strangely enough, the word is getting around, and rightly so.

The 1974 360 Bultaco is dubbed the "Jim Pomeroy Replica." The new goodies would take a full two pages to list. We'll just go over the ones that we feel are the most important and outstanding.

First on the list is the weight. With the chrome moly frame, handlebars and swingarm, the 360 weight is a meager 204 lbs. Tack on about four pounds for enough gas to run a 45 minute moto and you still have one very light machine. Although it is below the required FIM weight rules for international competition, Bultaco feels that the bike is designed for the local riders, not the top 10 professionals. Thus you get an ultra lightweight good handling bike that proves the theory that you can make a production machine exceptionally light. Now the other companies will have to come up with another excuse.

One of the first things you'll notice about the machine is that it lacks the brochure goodies. You know the type of things. Those little options that the advertising department feels are necessary to make the machine seem racey or pretty. Bultaco canned all those extra gimmicks and set about to make a race bike. Just a race bike.

The folks who make Akronts finally got their head straight and began manufacturing shoulderless rims. Although they are a little behind the rest of the rim market, we feel that their design and idea is one of the best. Bultaco uses the new shoulderless rims for both wheels of their machine, making it unnecessary to change this item when you purchase a box stocker. They have also incorporated those small sheet metal screws to hold the tire in place.

Laced to the front Akront is the Bultaco brake that made its debut last year. It's one of the better brakes on the market, very similar to that of the 250 Husky and Yamaha motocrossers. Brad Lackey ran one on his machine when he was still riding for Kawasaki. If it's good enough for Brad . . .

Bultacos have always been plagued by a case of post nasal drip in the fork department. It was getting to become sort of a tradition, like leaky Triumph cases. If your Bul's forks didn't leak, you hadn't ridden it enough. Now the Spanish firm has broken with tradition and installed a new set of fork seals. During our testing period, an accumulated 23 hours on the machine, we watched them with eagle eyes. So far they haven't leaked. But the proof will come in a couple more months of thrashing. We feel that they'll stay high and dry, for a change.

Many of the more safety conscious riders always worried about the steel number plate on the front. Some even had bad dreams about going over the handlebars, only to have a quick appendectomy performed by your bike. You can rest easy now guys. The new plate is plastic.

Not just an update, the new 360 is an entirely new Bultaco.

Moving right along, there's all kinds of fancy new doo-dads on the handlebars. First off the bars themselves are new. Chrome moly has been incorporated and the bars come on an extra wide scale. If you're a gorilla you can leave them the way they are. If you're a normal sized motocrosser, you can cut the ends down a bit and have a custom contoured ride.

All the cables that converge on the bars are now one piece units. No more nonsense of fidgeting with the little clamp and ball garbage. They now come with normal everyday cables. Nylon lined too!

Also up there you'll note that the throttle assembly has been altered to allow the owner to change cables with a minimum of trouble. The old one pops out and the new one jumps in its place. Easy as pie. Changing cables is no longer a Saturday afternoon chore. Both clutch and front brake levers are covered with new lever covers, keeping the mud of the course out of your controls. There's also a kill button floating around up there, just in case.

Remember last year when so many riders were seeing chiropractors about tailbone problems, due to the famous "Bultaco Hump?" Used to be that a large, rather obscene lump was sitting under the thin padding of the seat. It was situated right in the wrong place. The new Buls have cured that, with the

BULTACO
BULTACO
LTACO
POPULAR
CYCLING

Bultaco never has gone out for frills. The new 360 engine is plain looking, but sure delivers the punch when you give the throttle some running room. Double firing plugs are now standard.

You can almost bury the competition with this 4.50 Pirelli. It aids in cornering by allowing the rider to throw the machine over radically and remain in the seat. Cuts down on slides, though.

elimination of the lump and a better padded seat. No more grimacing and groans over the rough whoop-dee-doos.

The gas tank has a couple of new colors this year. It's still red, but has a brilliant yellowish-orange stripe. The 250 is blue. One minor aggravation with the gas cap was the overabundance of revolutions it took to remove the cap. It seems as though Bultaco went bananas on threads. You could never make a quick pit stop with it.

Under the tank lies one of the biggest inprovements on the Mk VII. From the motoplat ignition to the new chamber dimensions, the engine is all new. The porting design and applications have been changed. Instead of the usual configuration, there are now two transfers on each side of the cylinder wall. This coupled with the heavier and wider crankshaft makes the powerband smoother and more applicable to motocrossing.

One of the few external alterations visible is that now the Bul fires two plugs through the Motoplat ignition instead of the usual one plug. Two plugs should give you better fire and longer life for both your plugs and your running time plus less chance of oil fouling them.

Also new is the head steady mounting plate. The small plate that runs from the cylinder head to the frame is now one piece and is custom fitted to each motorcycle. It allows for easier repair and less hassle in taking out the engine. If for some reason you have to replace this part, you'll have to drill the holes yourself. They're all one-offs.

It seems that Bultaco is listening to their riders, mainly Jim Pomeroy. Jim can take a lot of credit for the production of both the 250 and 360 Buls. He has been working with the Telesco shocks for some time now and claims that they are just as good if not better than Konis. The shocks have a full five inches of travel, and don't seem to fade during those longer motos.

The air filter is top quality, a Twin Air element. These filters have been used by such top pros as Roger DeCoster, Willi Bauer and Joel Robert. When it wears out you can get a replacement from JT Racing. They're the only distributor, as of this writing, who carry these fine units.

Fiberglass fenders and side covers have always been the downfall of Bultaco riders. Either you invested in a resin company or made friends with a surfboard repair freak. Seems that the little devils would shatter in a stiff wind. Now many of the components are good old plastic. The side covers (number plates, etc.) are completely plastic. The fenders are a bit of a compromise, being about half plastic and half fiberglass. Oh well, it's the thought that counts. You couldn't expect Bultaco to throw out all those old molds and all that glass could you?

The traction end of the motorcycle is one of the most important parts and Bultaco put a lot of thought and time into testing and evaluating tires. What they came up with for a final product could probably pull your favorite tractor around for a while. The enormous 450 Pirelli not only gives you ample traction, but is six pounds lighter than the tire offered last year. This traction grabber not only works down the straights under hard acceleration, but helps in the corners and mud. Bultaco really knows where their knobbies are at.

The chain guide and rear brake anchor have both been beefed up while at the same time lightened slightly. On the top of the swing arm there is a small patch of rubber, to prevent your Bultaco from sawing the swingarm in half from a loose chain.

Exhaust pipes have been the plague of many motorcycle manufacturers. Either they look terrible and last forever or they look really sexy and fall off the first time you start the bike. The Mk VII has both an eye pleasing look to the exhaust pipe and it lasts for quite a while. Replacement packs for the silencer are available from your local Bul dealer. We suggest you get some glass from your friendly packing firm and do it yourself. It's cheaper too.

With all these new improvements and drastic changes, you'd think that the bike is easier to ride. It definitely is. The only problem that we had was starting the machine. A few times after you tickled the carb and got it running it would mysteriously run out of gas. It didn't load up, just ran out of gas. We suspiciously eyed the in-line fuel filter that someone had stuck on the machine. This little gem doesn't come stock, but when we took it off, we didn't have any problems getting the bike to run. Shows you just how well some of those fuel filters really work.

The biggest problem for a couple of our test riders was the adaptation of the right side shifter. After riding motorcycles with the left side shifter for years, it gets a little difficult to switch over immediately. It made for a few slow laps until your feet got coordinated with your brain to make the bike shift and stop. More than one berm and corner were overshot Ever tried to downshift your brake lever?

One of the first habits you'll notice on the new Bultaco is that it doesn't like to rev high and put out gobs of horses. To achieve the best traction and ground to horsepower ratio, you had to dial on the power from the low end and shift about mid-range. When operated in this way the machine cooperates beautifully. Try to treat it like a nervous 125 and the competition will only laugh as they go streaking past.

Contrary to popular belief, all Bultacos are not natural sliders. This particular machine wouldn't cooperate with you on any type of flat wide sweeper. With the large tire on the rear and the smaller skinnier gas tank you have a tendency to treat it like a 125 Elsinore. Up on the tank, flick it sideways and it wants to track through the corner like a train on rails.

Some of this is attributed to the Pirelli on the traction end. It is such a good tire that it allows only a minimum of wheelspin, thus you won't get those fantastic crossed up full lock slides that look so great on the photos. What you do achieve is some fast cornering and good traction out of the hole.

The Telesco shocks really came through for the 360 Bultaco, especially on the rougher sections of the course. They allow the rear end to track straight and true while the normal motorcycle is having difficulty staying in the groove. The abundance of low end power makes for a smoother and more relaxing ride, giving the rider a little more time on the bike before fatigue begins to set in.

Brakes were superb, but did require a

rather heavy hand to apply them. They're not the average power type of brakes that you find on so many Japanese machines. If you wanted the brakes, they were there, but you had to use liberal force to get them to cooperate. Even with heavy use, they never faded.

This Bultaco handles much like a 125, thus you find yourself cutting quite a few corners tighter than you would normally do on an Open class machine. If you can get in under the competition, you should have no problems getting into the straightaway and away before they know you're there. The light weight of the machine allows you to place it just about anywhere, while the rear tire helps to keep it there.

There's always the flashy type of riders around, and for those of you who are addicted to wheelies, the new Bul might be right up your alley. With the abundance of low end torque and the confident rev of the engine, you should be able to master those straightaway length wheelies in no time. We suspect that quite a few of the local photographers will be watching for this bike out at the races because it pulls such nice controlled wheelies. Makes for some neat photos.

Overall the new 360 Bultaco isn't really an update of the old Bultaco. It's an entirely different machine. It's light, corners incredibly fast, and has enough traction and power to keep anyone happy. For a price tag of $1395 it isn't the most expensive open class mount on the market. But it might just be the best buy for the money.

Looking all the world like its little brother the 125 Pursang, the new Bul is light, agile and ready to race out of the crate. Gas it like a big bore and handle it like a 125.

Mitch Mayes demonstrates that you can bottom out the exceptional suspension, but only after a monstrous jump.

BULTACO 360 MK-7

Max. Pts.	NUMERICAL EVALUATION	
10	Power	9
10	Powerband	9
10	Acceleration	10
10	Transmission	
	(5) Ratios	5
	(5) Operation	4
10	Suspension	
	(5) Front	5
	(5) Rear	5
10	Brakes	
	(5) Rear	4
	(5) Rear	4
10	General Handling	10
30	Miscellanea	
	(5) Starting	4
	(5) Rider Comfort	5
	(5) Quality of Craftsmanship	4
	(5) Riding Maneuverability	5
	(5) Tires	5
	(5) Noise Level	4
100 pts.	Overall Rating	92 pts.

Suggested Retail Price: $1445

ENGINE

Engine type	2-S, sgl. pist.-prt.
Bore and stroke, mm	83.7 x 64
Displacement, cc	352
Horsepower/rpm (claimed)	n.a.
Torque/rpm (claimed)	n.a.
Compression ratio	7.5:1
Air filtration	twin air
Carburetion	36 Amal
Lubrication	w/fuel
Ignition	Motoplat

DRIVE TRAIN

Transmission	five
Clutch type	wet, multi-disc
Primary drive	chain
Final drive ratio	13/46 3.538:1

CHASSIS

Chassis type	single downtube
Overall length, in.	82.7
Seat Height, in.	34
Peg height, in.	11.25
Ground clearance, in.	8.7
Wheelbase, in.	55
Weight, lbs. (as tested)	204
Tires, front	3.00 x 21
rear	4.50 x 18

BULTACO'S CHROME-MOLY 350 ALPINA

Handlebars determine where you sit on the seat: footrests are responsible for getting your knees into the grasshopper position.

Stand-up riding is almost correct for this 6-foot rider. Taller bars would allow him to bend less at the waist.

Shift lever is just an inch away from toe of size 11 boot. You've got to get off the peg to change gears.

Air filter is oil wetted foam type. Small pad is easy to service, adequate for 350 but doesn't surpress noise.

New speedo mount has eliminated Bultaco's "rotting rubberband" problem. Mounting plate is non-glare crinkle finish, trick.

350 ALPINA

Ever watched observed trials competition? Put you to sleep? Nobody goes fast, no danger, thrills or spills? No wonder you've never seen one, most other people haven't either. Observed trials seem to gather only those keenly interested in the sport.

And that makes us wonder how come so many of our motocross and other racing champions spent their first competitive years in observed trials competition. Even Jacky Ickyx, the Belgium car racing champion, won his country's observed trials title before he tried motocross, road racing, and eventually four wheels.

You could say observed trials is the primer to all forms of motor racing. It teaches you balance and how the application of power works. Concentration and determination, those two ingredients so vital to becoming a champion of anything, are the very crux of observed trials competition.

But you can't do it properly with just any old kind of bike; it takes a special one. Trials motorcycles practically turn within their own wheelbase, climb vertical walls from a dead stop, and wheelie all day. All they need is a rider of considerable skill.

Now Bultaco's Alpina is a derivative of that company's many-time World's Observed Trials Championship motorcycle. It's a bastard though, and came about because there's a definite need for that kind of mountain goat. This is how the Alp came to be:

Lots of observed trials bikes are sold that never see an observed trials. The guys are taking them off to the woods or up in the mountains to quietly ride trails and paths without so much as snapping a twig or spoiling the surface. The bikes are extremely light, making them ideal for mountaineering, and the engine's power to the rear wheel is as soft as a down pillow. Knobbies are never worn, trials tires do the job.

Problem with a pure trials bike for an all-day outing is a lack of fuel capacity (good for about 40 miles), too hard a saddle (trials riders stand up a lot), and the gear ratios are funny (trials bikes use the first three very close to each other).

Why not use an enduro bike instead, you say? Well they're heavy, don't turn too sharply, and most have the wrong kind of power. Some guys would rather bolt on a big gas tank, pad the seat and ride their trialer.

So along came the Alpina. Mister F.X. Bulto was the innovator (with the help of Sammy Miller) of the 250cc 2-stroke observed trials motorcycle. Mister Bulto is also a purist. He builds a Pursang for motocross, Matador for ISDT competition, the Astro for flat tracking, Sherpa Ts, and very fine road machines called Metrallas. Although sharing many of the same pieces, they are all specialized motorcycles with their own identity. Had Mister B known what the factory was up to in creating the Alpina it would have been squelched before that first fat-saddled, blue-tanked bike ever saw the outside of the adobe walls where it was assembled. "Sacrilege" he would shout, and the bike would have been dismantled, its pieces flung to the winds, with no chance of reconstruction.

Those early Alps were simply Sherpa Ts with Matador observed trials gear sets and a fuel tank/seat of its own, certainly useless for work. The first change was shelving the skinny trials rear shocks and mounting Matador units, the seat got fatter too.

After two years a new frame, glass, and triple clamps made it into a different kind of Alpina; they also added a 348 engine. The new Alp went from 230 to 246 pounds full of gas, gained height from a still fatter seat, and got big lights as it started leaning towards the street.

Now the Alp is firmly placed in the market, it's the bike for going where, if you had no other choice, you'd walk. Because it's not meant to be ridden fast, it has very soft suspension. Both wheels follow the terrain rather than hitting the high spots, allowing good control when everything gets gooey or bumpy or both.

Basic change on the new frame ('74) was an increase in trail. They did this by using MX triple clamps that bring the front axle closer to the centerline of the steering head. (The MXers have still more trail because their front axle is directly under the fork slider while Alpinas and Sherpa Ts have the axle in front of the slider.) It was taller, braced more rigidly, and weighed 16 pounds more.

'75 becomes the year of the "left side shift" Bultacos. This irks F.X. Bul-

Here it is folks, the shifter on the left. Aluminum arm has more bends than the letter "S"

27mm Amal carburetor is mounted on angle to allow "cooling room" between it and exhaust pipe.

Trailing chain tensioner has high walls so chain won't jump off. Aluminum nut on swing arm is filler for chain oiler.

Huge taillight is mounted on rubber chunk. Exit tubes in muffler are strange but sneaky tone is appreciated.

In profile, muffler looks like pregnant boomerang. New rear wheel has right side brake with reverse side capabilities.

to because he says if you stand on your left foot you stick out your right hand for balance. That means that with the throttle already at the right hand, the clutch has to be on your left so the shift lever must be for your right foot. It ain't no more.

Bulto relented begrudgingly (after a fight) to the new Federal regs, but with reservations. You can still have right foot shifting on your Bultaco. He left the splined shaft for a shift lever on the right side. The entire rear wheel flops over and the sprocket will bolt to the brake side. You only need to weld a tab on the swing arm for the brake anchor. Find the old-style shifter and brake lever and you've got it. Now you can stand on one foot and wave the other hand.

That wheel itself, both wheels in fact, are the most advanced and sophisticated found on a production off-road motorcycle. Although hard-chrome plating aluminum friction surfaces for long life has been possible for many years, we've only seen it on rare occasions in cylinder bores of motorcycles. Aerospace has been into it for decades.

The wheel hubs of both wheels on the Bultaco are eggshell-thin aluminum forgings. The area that makes contact with the brake shoes is hard-chrome plated, a .005-inch coating (about the thickness of one of Bulto's whiskers) that rates near tungsten for hardness. The advantage is lightness. Normally there would be a thick cast iron or steel liner sleeve in the aluminim hub for the brake shoes to rub against.

A two-piece bolt-up assembly makes up the rear hub. With the sprocket carrier opposite the brake drum, the need for different length spokes has vanished. The Alpina rear wheel is the same as found on Pursangs, 8-gauge spokes and all.

The conical hub on front is shared with the Matador. Non-lipped Akront hardened rims, stainless steel spokes and aluminum alloy nipples help in making these the very lightest weight wheels in the industry. Patterned after the tiny Pursang front brake, the Alp's is larger and fitted with an internal speedometer gear set.

Bultacos are surrounded with innovations for improvement. These speedometer gears used to be of cast iron, they needed lubrication and a working area free of foreign matter. Sometimes a machine chip from some previous machining operation would get drawn in by the grease and the gears would grind to a halt. Currently a nylon ring gear has been substituted to ride against a metal pinion without lubrication. So far the system is pretty successful.

But not all these progressive ideas from the Barcelona factory have worked out. Second generation Alpinas were using a double O-ringed cork for a gas tank cap. The thing took Hercules fingers to remove and was just as difficult to force back into the hole from which it came. Solution; return to the old screw-type.

Actually, the best step forward made with the '75 model is the switch to a chrome moly steel tube frame. (That is if you're not freak about which foot you like to change gears with.) Dimensionally the same as last year's "longer-taller" Alp, the change in material saved them a total of 8 pounds on our balance beam scale. The '75 works better for still another reason, its got a longer wheelbase because the swing arm is longer. The bike rides a little softer and turns a little slower.

Hanging in there with this "slower" theme is a one-third-turn throttle, several pounds added to the flywheels, a tiny 27mm carburetor, and "pregnant boomerang" after-exhaust system. Together these are the things that make you think the engine is really an electric motor hiding inside motorcycle engine cases and castings. The rubber-housed throttle is an electric rheostat. Power, although there's not much of it, is that smooth and controllable. If you want to learn to do wheelies, this is the bike to use.

That "thing" on the end of the exhaust system does come off without too much effort. Without it the Alp makes noise (it didn't before) and actually runs a little stronger. There is no exhaust tuning in these engines. Small carburetor and mild port-timed observed trials-type horsepower doesn't require expansion chambers. There is a glass pack muffler ahead of the "boomerang" that keeps the note down for a few hours running until the glass has "packed." Then it's just plain obtrusive.

Continued on next page

It comes pretty well set up for enduros. Rubber covers around the two hand levers and full protection to the throttle cable indicate this bike could wallow in the mud with the best of 'em. It's certainly got the kind of power to get you out of those slimy spots, and should *crawl* up a greasy hillside as well. The tenth-resettable Veglia-Bressel speedometer will give you an accurate account of the miles you've slithered.

They've done a nice job with the air filter too. Hidden under an orange cover of the orange and blue color scheme, is a genuine Twin Air foam pad. Its well sealed and requires incoming air to rush in from under the seat, a feature that makes the new Alp particularly well suited for splashing about in streams and shallow pools.

Wide, deeply valanced fenders will keep the junk off you, even the mud flap up front helps. The aluminum front fender is held by three new huge-head screws that grunch in the tire tread when you bottom the forks. And the springs are so soft it's not difficult to do. A couple Bultaco fork spring spacers should have been factory fitted.

The rear fender is stainless steel and mounts the rubber bracket to which the large taillight is screwed. This practice of rubber mounting the lights, including the headlight, doesn't make a whole lot of sense to me. I had to ride the Alp after dark out in the boonies and that wiggly headlight never stayed in one place long enough to spot anything.

The new side stand position is worth mentioning; it's on the swing arm like the new Sherpa Ts. It's got a new right-side engine case and the cross between a Matador and Sherpa carries rollers instead of bushings on the transmission mainshaft, like the Pursang. So the saga of the Alp will continue to unfold every six months, and from it a wholly new concept in off-road riding. A serious bike for the serious non-racer, a mountain machine for mountain men. A good vehicle for semi-skilled riders to break into off-road, and the kind of machine an expert never gets tired of because of the bike's capacity to make him look even better wherever he goes. •

BULTACO 350 ALPINA

TEST BIKE

Engine serial	11800247
Base price as tested	$1395.00
Factory warranty	90 day mfg. defect

ENGINE

Type	2-stroke piston port
Displacement	348cc
Bore x stroke	87mm x 64mm
Claimed HP @ RPM	N.A.
Claimed torque @ RPM	N.A.
Compression ratio	10:1
Lubrication system	Oil mist
Carburetion	27mm Amal
Air filter	Oil wetted foam
Ignition system	Magneto. points. coil
Electrical system	Alternator. no batt.
Starting	Left foot. indirect
Exhaust	Muffler & spark arrestor

DRIVE TRAIN

Primary/ratio	2.37:1 Duplex chain
Clutch	Wet-all steel multi-plate
Transmission/shift	Leftside
Gear ratios	(Internal) 1st .287:1; 2nd .442:1; 3rd .625:1; 4th 821:1; 5th 1:1
Final drive/ratio (sprocket teeth)	3.23:1 (13/42)

CHASSIS AND SUSPENSION

Frame	Single downtube. one-piece chrome moly steel welded structure
Suspension, front	Betor 2-way damped 7" travel
rear	Betor 2-way damped. 5 position adjustable spring
Turn radius	N.A.
Brakes, front	SLS-internal expanding
rear	SLS-internal expanding
Tires, front	3.00x21 knobby
rear	4.00x18 knobby
Rim locks, front/rear	Rim screws

WEIGHTS AND CAPACITIES

Weight, wet, unladen	238
Allowable gross weight	N.A.
Fuel capacity	2.8 gals.
Engine oil	in fuel

STANDARD EQUIPMENT

Speedometer	10th resettable
Tools	Yes
Stands	R/H side
Passenger provisions	No

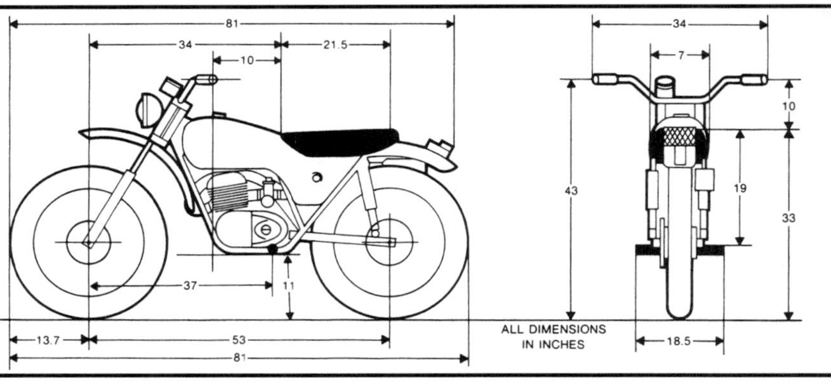

BULTACO 250 PURSANG

Yet Another Superlight, But This Time With Radical Suspension

Cycle World Road Test

■ KNOWING A LITTLE of how Spaniards think is helpful when trying to understand why they do things the way they do. Spain's people are very proud. They know that their country isn't the richest one around and that they don't have as many resources as many other countries do, but they know how to take advantage of what they do have. . .both in the area of resources and in the area of technology. The latter has been the strength behind their successful motorcycles. The technology is of the homemade variety. They figure things out for themselves, and it works. No megabucks, electronic thingamawidget telling them how to design motorcycles.

Their machines have come about through many years of trial and error. They love motorcycles, not just as a source of income, but simply as motorcycles. They take pride in their accomplishments and, as long as they continue to produce competition machinery as fine as this newest Bultaco, they have every right to.

This is Spain's first attempt at mass-production of a FMS (forward-mounted shocks) motocrosser. At least it's the first to hit our shores. Montesa has just released its slant-shock VR, of which CYCLE WORLD will have the first test, but which won't arrive for another month or so. Until then, the Bultaco will do fine. Just fine.

The 250 Bul carries the same paint scheme that was introduced on last year's model. Blue with white stripes. At first it seemed a bit odd and difficult to get accustomed to, since we've always known Bultacos to have red paint somewhere, even if it's only a touch. But after a while, you begin to accept the blue Bul and the color never bothers you again. In fact, it sort of starts to grow on you.

The Mk VIII is a little heavier than last year's Mk VII. At 206 lb. dry, it carries four pounds more. While some of the modifications on the Bul make certain components lighter than before (such as the new right side cover that more closely follows the contour of the internals), the FMS conversion, along with a new hub, makes up the difference in weight. In order to maintain as much swinging arm rigidity as possible, the full length of the arm has been gusseted along its lower side with a piece of curved steel tubing. The gusset, swinging arm and entire frame are made from chrome moly.

The rear hub is new in that it is reversible. The brake drum can be placed either on the same side as the sprocket or on the opposite side. This, of course, works in conjunction with the new shift shaft that runs completely through the cases, allowing the rider his choice as to which side to shift on and which side to place the brake pedal on. As you buy it, the machine comes with a left-side shift and right-side brake. One of the reasons for the change was that some Bultaco people felt that sales were being lost to customers who were graduating from a left-shift Japanese bike and didn't want to have to relearn shifting reflexes.

This arrangement is most noteworthy because it gives the rider a choice. It doesn't alienate long-time Bul riders, while at the same time appealing to those who might otherwise deny themselves the pleasure of owning a Pursang. Also, riders who take the time to become proficient shifters with either foot can take advantage of the adjustability of the machine to set it up for a particular track, depending on which way it circles and where the most troublesome corner approaches and exits might be.

No matter which side you choose, the rear brake works

very well. That's more than can be said for the front one, though. At first, ours hardly worked at all. Little by little it got better as the hours were piled onto the machine, but it never stopped with what we would consider sufficient strength. We understand that part of the problem is that in order to keep things light, Bultaco discarded the brake drum lining in its Dural hub and uses a metallic spray finish instead. The finish doesn't peel or anything like that, it just doesn't create enough friction.

The attempt to keep things light did succeed. We've never ridden a machine, any machine, that was so totally responsive to body position and movement as the Mk VIII was. Anything you wanted to do you could do on this bike. Wheelie forever, slide like Dave Aldana, take the inside line, the outside line, carom off berms, make Saturn V leaps off jumps, play DeCoster in mid-flight, or, even better yet, do some of those patented Rex Staten double cross-ups. The Bul won't let you down. And even if you don't get completely straightened up from your antics before you land, the Bul lands softly and straightens up immediately. It's uncanny.

When the playing around is finished and it's time to take care of business, the real fun begins. Of course, with the mediocre front brake, you'll have to work extra hard to beat anyone going into corners, but getting through and smokin' out is a blast. Again, the incredibly light feel allows you to toss the bike around and take any line you like. The Pirelli tires stick well, even on some of the dry tracks to which we took our test bike. The monstrous 4.50 tire we've always liked, but the fact that the skittery 3.00-21 up front didn't wash out really surprised us. Sure speaks well of Bultaco geometry.

To get the front end up you must merely think the word "up" and there it is. Almost like magic. Yet when you want the front end to stay down and track straight, it does that just as automatically. At times, it's as though the machine can read your mind. A light front end alone can't just get itself up in the air, though. You need good strong power to get it lofted, and that's just what the engine delivers.

Power starts building in the lower rpm range where many 250s can't even run. Only the VR Montesa and the Ossa Phantom have exhibited similar power characteristics. Mid-range was the Bul's strongest area of power delivery, while the top end wasn't too shabby either. If you only want to talk about outright screaming horsepower, the Bul is down a little to most other bikes in its class, and even more to engines like the Can-Am 250. But the mid-range is particularly strong, and that's where the torque peak lies. When you keep the engine running in this range, the bike rockets out of corners. It doesn't erupt or explode with wheel-spinning roostertails of dust and debris, but it grabs hold of the ground and comes out like good ol' greased lightnin'. For the backyard tuning whizzes who just drool at the thought of attacking a stock barrel with their roto-rooters, there's room for improvement, but you'd be messing up a really great powerband just for an extra pony or two.

The 36mm Amal carburetor appears, on paper, to be more than the bike needs. Yet, the engine starts on a few kicks at worst and runs clearly throughout the range. No doubt the pointless Motoplat ignition helps.

Suspension-wise, the story of the day sits aft of the engine. Offering just a tad more than six inches of travel at the rear axle, the Telesco shocks—which, surprisingly, are not of the gas/oil variety—mount in a perfectly vertical position. They work great and didn't fade during the weeks we had the bike.

The frame on the Bul hasn't been modified at all for the FMS. Just the swinging arm has. The chassis is the same as last year's. Performance of the rear end is super. A little harsh for just putting around, the set-up really comes into its own when the throttle is dialed open. The whole suspension system is great. The forks react to all the pounding you can dish out by responding perfectly. After a while, the left fork leg on our test bike (which, by the way, came to us courtesy of AAA Cycles in Santa Ana, since the Bultaco distributor was completely sold out), began to leak. The trouble was traced to a flapper valve on the damper rod that was kind of hosed, causing extreme hydraulic pressure and the resulting leak. There *was* one problem that the rear suspension caused, and there's little anyone can do about it.

This is our major complaint with the bike. It is far too tall for most people. The seat height crests the 35-in. barrier; and even six-footers on our staff found sitting astride the Bul a tiptoe situation. Just to see what would happen to someone decidedly shorter, we asked a guy who was 5-ft. 4-in. to sit on the bike and he had one heck of a tough time. To the racer, the only time that the tallish stance of the machine is a problem, is on the starting line. Shorter riders might need blocks to stand on as they steady their machine, waiting for the gate to drop. The thought of that seems funny, but we don't see any other way.

Once underway, everything fits fine. Riders in a large range of sizes can feel at home with the positioning of the controls. The handlebars may be too wide for some, but that isn't a serious problem and can be easily corrected.

As we're sure you can tell, we liked the new Pursang immensely. It has a great number of features that, in addition to its performance, make it desirable. There are non-ridged Akront rims, complete with sheet-metal-screw-type pins in the rims, thereby eliminating the need for rim locks; there are the best footpegs in the industry, good grips, aluminum levers, and a comfortable saddle. You can also add to the list a foam filter >

250 PURSANG

element in a well-protected airbox, a chain guide and brake anchor arm made from Dural, and even replaceable fiberglass packing for the quiet downpipe. But the details are not entirely without fault. We hoped the fiberglass fenders would be a thing of the past this year. They aren't. The chain stretches readily; and we wish the fuel tank were plastic instead of fiberglass. A little more volume wouldn't hurt either.

Overall, the newest Bultaco is a true delight. It isn't often that we come across a machine that allows us a chance to writ as laudatory a road test as we have here, but the machin deserves every word of praise that it has received. It is fas light, handles superbly and truly comes ready to race. Th Spaniards have really done themselves proud on this one.

But, above all else, the Bultaco Pursang Mk VIII is a ston gas to ride. Pure, undiluted fun. And if you haven't foun motocross to be much fun lately, this one'll put you right bac in the game.

PARTS PRICING

Cylinder	$121.47
Cylinder Sleeve	37.34
Cylinder Head	54.68
Piston Kit	33.84
(1) Set Rings	11.20
Rear Shocks (each)	29.76
Front Hub (comp)	91.88
Rear Hub (comp)	179.80
Spokes (each)	0.60
Wheel Rims (bare each)	39.80
Drive Chain (standard)	19.50
Front Fender	9.95
Rear Fender	14.95
Clutch & Brake Levers (each)	8.00-10.00 est.
Clutch Cable	4.95
Throttle Cable	4.95
Brake Cables	4.95
Ignition Parts	
Sealed Unit Type (comp)	109.67
Carburetor	74.43
Crankshaft	138.10
Connecting Rod Kit	41.60
Shift Lever	12.00-13.00 est.
Brake Pedal	12.00-13.00 est.

BULTACO 250 PURSANG

SPECIFICATIONS

List price	$1495
Suspension, front	telescopic fork
Suspension, rear	swinging arm
Tire, front	3.00-21
Tire, rear	4.50-18
Engine, type	piston-port, two-stroke Single
Bore x stroke, in., mm	2.83 x 2.36, 72 x 60
Piston displacement, cu. in., cc	14.9, 244.29
Compression ratio	12.0:1
Claimed bhp @ rpm	N.A.
Claimed torque @ rpm lb.-ft.	N.A.
Piston speed @ rpm ft./min.	3736
Carburetion	36 mm Amal
Ignition	Motoplat pointless
Oil system	pre-mix
Oil capacity, pt.	1.3
Fuel capacity, U.S. gal.	1.8
Recommended fuel	premium
Starting system	kick, folding crank
Air filtration	oil-wetted foam

POWER TRANSMISSION

Clutch	multi-plate, wet
Primary drive	dual-row chain
Final drive	525 single-row chain
Gear ratios, overall:1	
5th	9.10
4th	10.92
3rd	13.56
2nd	17.75
1st	24.12

DIMENSIONS

Wheelbase, in.	55.5
Seat height, in.	35.0
Seat width, in.	7.5
Handlebar width, in.	34.0
Footpeg height, in.	13.0
Ground clearance, in.	10.0
Curb weight (w/half-tank fuel), lb.	215
Weight bias, front/rear, percent	45/55

Photography: Randy Papke

Bultaco 360 Astro

A Professional Machine For The Professional Rider

There's a saying among pilots that goes something like this; "A plane that looks good usually flies good." Of course that's not always true, but more often than not it is. In many ways the same could be said about motorcycles. If it looks good it usually works good. A Maico is a good example; it certainly doesn't look pretty but it looks efficient, and as we all know it does get the job done.

Another motorcycle that not only looks efficient but that can be called beautiful is the Bultaco Astro. This bike is meant primarily for professional short-track racing, but with a few minor changes and a front brake attached, it also doubles as a fine TT rig. When you first look at the Astro you get that gut feeling that says here's a bike that can make it, and you're not wrong.

In the past the 250cc Astro was known as a strong bike with a short fuse. It got this reputation for a good reason—it wasn't the most reliable motorcycle ever put on the race track. Most knowledgeable Bultaco racers went through the engine after every race. If they didn't it wouldn't make the next complete night of racing. That's not true anymore however, ever since Bultaco went to the 359cc engine, which is legal under AMA rules. The machine has become super reliable and it's still ungodly fast.

A short-track machine is completely different from a motocrosser. A track that a short-track rider would call rough, a motocrosser would think ridiculously smooth. And probably the biggest difference is that during a short-track race the bike only has to turn left. It's kinda like the Indy cars, if you have to turn right you're in deep trouble.

Because the short-track machine has only these relatively simple tasks to perform, it's much less complicated than the average motocross motorcycle. It doesn't have a front wheel brake, instead a spool hub is used to hold down the weight. Long fork travel isn't wanted, so the forks used can be considered a cross between motocross and road racing units. Nineteen-inch rims are used both front and rear, and the only brake on the bike is a hydraulic disc on the rear. The rear wheel itself features a quick-change hub for the sprockets, and the gas tank capacity is very small, just over one gallon. The handlebars, instead of being thought of as straight and flat, like motocross bars, sweep upward more and come back to the rider somewhat in the manner of wheelbarrow handles. When you get right down to it, the only similarity between a short tracker and a motocrosser is that they both run on dirt, one on the smooth, groomed short track, and the other over the roughest ground the track builder can find.

One item a track machine must have over a motocrosser is horsepower, maximum all-out muscle. Whereas the motocross bike uses all of the torque band during a race, the tracker only uses power crammed into the upper end of the rev range. In the Bultaco line for example, the Pursang only revs to about 7000 rpm. On the other hand, the Astro revs slightly over 9000 revs and makes its maximum power between 6000 and 9000 rpm.

Since maximum power is needed for short tracking, the engine inside is somewhat different than the motocross Pursang. Displacement of the Pursang is 363cc, while in order to stay within the AMA rules the Astro displaces only 359cc. The stroke in the two engines is the same, 64mm, but instead of using the same piston Bultaco has utilized the third oversized Alpina piston to give a bore of 84.5mm. Another difference is the 22mm crankpin, even though the latest Pursangs also use this. Part of the reason for the troubles in the past was that the crank throws, or internal

A spool hub is standard at the front. If the bike is used for TT racing, then a Pursang hub and wheel are used.

In standard trim the Astro is one of the prettiest competition motorcycles you can buy.

(Below) Being a short tracker, both the shift lever and the rear brake lever are on the same side. This would have to be changed if the bike is to be used on a TT track. Everything is designed so as to make this task easy.

The old tried-and-true 36mm Amal concentric carb is used. It works just fine, providing a rev range without flat spots.

flywheels, shifted under maximum revs. The larger crankpin was used to cure this, and we're here to tell you that it does. During our day of testing this bike, we put almost five hours on the bike. That would be comparable to almost two years of racing short track. Usually at a short-track program a rider will run practice, time trials, a heat race, maybe a semi-main, and then if he made the program the main event. All of this would be less than 15 minutes of time on the engine. You can drop the short fuse tag from the Astro, it isn't true anymore.

Instead of using the newer Mark Two Amal carburetor the Astro still comes with the concentric unit. They use this carb because it flows air at the top end of the rev range better than just about any carb ever built. It's 36mm, the same size as that found on the Pursang.

The pulling power of the Astro comes somewhat as a surprise. Because it does rev so high you might think it to be a fairly pipey engine. Not so. The engine pulls cleanly up onto the pipe with no flat spots in the rev range. We were surprised until we found out that the bike comes with the same external flywheel as the Frontera. It's big and it's heavy, but it smooths out the power pulses and gives the added plus of helping to carry the bike into a corner after the rider has shut off the gas. Not all track riders like the heavy unit, so it's possible to put on the lighter flywheel that the early '75 Pursangs came with. The later ones have the same flywheel as the Astro and Frontera.

Because the Astro engine revs so much higher than the Pursang you'd expect to find differences in both the expansion chamber and the porting. In both cases you'd be right. In the Astro the intake port is lower and the exhaust port higher. The Astro's pipe has more volume than that found on the Pursang, but the latest Pursang pipe is closer to that of the Astro than the earlier '75 Pursang.

Other than these differences, the engines are alike. The internal gear ratios, clutch and primary drive, engine cases and head are all the same.

When you start looking at the suspension you find components specifically designed for short-track work. Instead of the long fork legs found on the motocross bike, you find short forks that would look more at home on a road racing machine. The forks have four and a half inches of travel, more than enough for short-track racing. Forks are of course Betors, you find them on all Bultacos.

At the rear are special short-travel Betor shocks instead of the Telesos found on motocross machines. They only let the rear wheel travel slightly over four inches, again just what's necessary for short-track racing. Suspension at both ends is somewhat stiffer than that found on a motocrosss machine.

Tires for short-track racing are a science in themselves. Consequently it's hard for a company in Spain to put the proper rubber on a motorcycle. When you buy the Astro it comes with Pirellis mounted, and either Carlyles or Goodyears as spares. It's more than likely that none of these three are what the rider will need for a given track so he'll carry others with him. He also might change tires two or three times a night as the track conditions vary from race to race. Interestingly, the second set of tires are placed in the crate in this country when

On a groove track, the bike will track comfortably if it's wearing the right tires.

The left-side peg just dangles when the bike is used for oval racing. In TT competition, this peg is removed and a regular spring-loaded peg is put in its place.

A compression release isn't standard, but the head is drilled so that the rider can mount one if he so desires. Most short-track riders use one to help in braking the machine.

the bike hits Virginia Beach, the location of the U.S. distributor. That's also when the disc brake meets the bike for the first time. The disc brake setup in fact was more or less designed in this country. Two Bultaco team riders, Guy McClure and Roger Crump, experimented around and found the present arrangement to be the best. Instead of shipping the American-made unit to Spain to be put on the bike there, Bultaco waits and puts it in the crate here.

Because track bikes only turn left, the brake pedal and the shift lever are both mounted on the right side of the bike. The brake lever sits very low to the ground and the shift lever is above it. The footpeg on the right side of the bike can be relocated in different positions so that the rider can get his weight where he wants it for different track conditions. The left peg sits higher on the frame and toward the rear of the main frame section. It's only used to keep the rider's foot off the ground between corners. Remember, this bike is meant to do its job when turning left.

When it comes to bikes that turn left for a living, you're talking about a lot of money. Not many manufacturers build short-track bikes. We know only of Bultaco, otherwise a rider has to make his own. That means buying an engine, frame, forks and shocks, and all the rest of the running gear. We've been told by those who know that it costs between $2000 and $2500 to put a tracker together. On the other hand, you can spend $1795 and buy the Astro, truly a machine built with the professional rider in mind.

Riding a track bike is different, much different than a motocrosser. Setting the bike up for a corner goes against everything you've ever learned. To get the bike into a slide you kinda take a big step to the left and turn on the gas. This gets the bike sliding, and then you get off the throttle, all the while keeping the bike in the slide. This is the best and fastest way to slow down. As soon as you get down to cornering speed you climb back on the gas and steer with the throttle. Coming out of the corner you keep feeding in the gas gently so that the rear tire doesn't break completely loose. As it is, the Astro has the kind of power that will spin the wheel from one corner to the next. Our test track, about 1/8-mile long, could be gone around with the rear wheel never getting a solid bite if the rider had a heavy throttle hand. But the interesting thing is that with Bultacos being some of the finest sliding bikes ever built, the Astro likes to be "blue grooved" if the track allows it. Much of this depends on the rider's choice of tires, but it is interesting that the bike will hold a groove if it's there.

On the East Coast there are many sportsman short-track races. The 360 Astro can be used there, and for the 250 class, Bultaco also makes a 250 Astro. It costs exactly $100 less than the 360, and it just about rules supreme against the other quarter-liter bikes. Both versions can be made to shift on either side the rider prefers, and of course, a front wheel brake can be added for TT racing.

The Astro is meant for smooth-track racing; a motocross course would easily defeat it. But in what it does there are few bikes that can keep it in sight, say nothing about beating it. Plus, the bike you had to build to beat it probably cost you $600 to $800 more than the Astro. You pays your money and you takes your chances, just make sure you pick a good bike.

Max. Pts.	NUMERICAL EVALUATION	
10	Power	10
10	Powerband	10
10	Acceleration	10
10	Transmission	
	(5) Ratios	5
	(5) Operation	5
10	Suspension	
	(5) Front	5
	(5) Rear	5
10	Brakes	
	(5) Front	Not Applicable
	(5) Rear	4
10	General Handling	9
30	Miscellanea	
	(5) Starting	4
	(5) Rider Comfort	5
	(5) Quality of Craftsmanship	4
	(5) Riding Maneuverability	5
	(5) Tires	4
	(5) Noise Level	5
100 pts.	Overall Rating	90 pts.

(Out of a possible 95 pts.)

BULTACO ASTRO

Suggested Retail Price: $1795.00

ENGINE

Engine type	single cylinder, 2/stroke
Horsepower/rpm	n.a.
Torque	n.a.
Bore and stroke	84.5x64
Displacement	359cc
Compression ratio	10.5
Carburetion	36mm Amal
Ignition	Femsa electronic

DRIVE TRAIN

Speeds	five
Clutch	wet, multi plate

CHASSIS

Length overall, in.	81
Wheelbase, in.	adjustable 50-54
Ground clearance, in.	5
Weight, lbs.	190
Frame type	single downtube
Tire size, front	4.00x19
rear	4.00x19
Brakes, front	none
rear	hydraulic disc

BULTACO SHERPA T 350

A Remarkable Series Of Improvements Keeps The T At The Front Of The Trials Bike Brigade.

Cycle World Trials Test

■ IT'S OFTEN VERY HARD for us to imagine how certain motorcycles can be improved upon from year to year. So impressive are they the first time around and in subsequent model years, that creating a superior version seems an impossible task. One such machine effects us this way year after year. It's Bultaco's Sherpa T trialer. We keep saying, "How are they ever going to top this?" And sure enough, pretty soon there's a new T and we're left standing there saying, "How are they ever going to top this?"

For 1976 we face the same situation again, because the new Bul trialer simply defies all odds of being improved upon. Styling, of course, is a matter of individual taste, but we doubt that we'd get much opposition if we declared the T a real beauty. The look is sculptured, with *narrow* as the theme. The frame is entirely chrome-moly and finished in Bultaco's traditional silver. Fuel capacity has been increased to 1.7 U.S. gal., yet the fiberglass tank retains an ultra-slim profile. Painted in high-gloss red with a silver accent band that matches the frame's color, the tank features a new, easy-grasp, vented plastic cap and an improved fuel petcock. No crossover tube is used, allowing tank removal with less hassle.

The front portion secures with a single 13mm bolt running through the gusset portion of the steering head, and is rubber-cushioned. The rear is held in place by the tiny seat assembly that snugs under a frame bracket and is tightened up at the rear with two 10mm bolts and self-locking aircraft-type nuts. In fact, self-locking nuts are used nearly throughout the machine.

Side panels are fiberglass and finished in red to match the tank. To help prevent scuffing from the rider's boots, the panels are indented and form fit to the frame rails, which have been narrowed this year. The left panel is actually an integral part of the new airbox assembly, while the right one removes with a single large screw to allow air filter servicing.

Made from fiberglass and drawing air from up high under the seat, the airbox is designed to be virtually watertight. A wire mesh screen keeps large debris from entering the chamber, over the top of which fits a thin foam element that is held in place by an aluminum bracket, which is pressure loaded by a spring that tightens down when the outer cover goes in place. A rubber lip seals the removable cover portion. Someone would have to work hard at drowning out this one in a water hazard.

As we mentioned before, the frame is new and manufactured from chrome-moly tubing. . .including the swinging arm. Tubing is smaller in diameter than what is normally found on Bultaco motocrossers and enduro machinery, because strength is less important and the weight factor is more critical. The frame, like before, is designed around a single toptube and downtube assembly. The downtube splits into an engine cradle where the tubing curls upward to form the rear triangulation sections that support the swinging arm, shock mounting points and bracketing for the seat and rear fender. Welded between the two lower frame rails is a perforated bash plate, providing a token amount of protection at best. A few companies make after-market skid plates for the T, Bultaco's own Miura Products being one. The Miura Products plate protects both left and right side engine cases, as well as the expensive-to-repair chrome-moly lower frame tubes. It's slippery plastic and super strong, cheap insurance at $14.95.

There is ample gusseting in the stress areas in the steering head and pivot point for the swinging arm; the tucked-in footpegs are also amply braced. Bultaco has succeeded in producing a lightweight, yet strong and efficient, frame for the purpose intended: observed trials competition. The steering head angle remains at 27 degrees, with 3 in. of front wheel trail.

Over the years there has been plenty of talk about proper tires for the serious trials competitor. And across the board, the most popular have been the British Dunlop two-plies with their flexible sidewalls and good gripping compound. But the tires have usually been hard to find for many, and some have even been fooled into purchasing the Japanese Dunlops, which are about as similar to the British versions as "Adam-12" is to *The Choirboys.* And then here comes Pirelli with its new two-ply nylon MT 13 Mototrial. We think it's every bit as good as the British Dunlop. . .and it's standard equipment on the new Sherpa T to boot!

The plastic injection-molding industry in Spain has much progress yet to make, and for this reason many of the Bultacos come with the long-familiar polished aluminum fenders. We all know that sooner or later the aluminum gets to looking pretty shabby from assorted prangs in the wilderness, yet we prefer them to the more brittle fiberglass pieces we sometimes see. The answer, of course, is an accessory fender by any one of many after-market companies. . .Preston Petty being about the best and most popular. With reasonable care (and luck) the aluminum fenders will last a while, but at replacement time plastic would be the way to go.

Both stock fenders keep a fair amount of goop off the rider. The front unit, which mounts close to the tire in normal trials fashion, has a rubber mud flap on the bottom to help keep the slop down. The low-mounted fender allows the rider to watch the placement of his tire in trials sections, which would be all but impossible with a motocross-type fender mounting. One problem that can crop up with the low fender, >

SHERPA T 350

however, is the jamming of the tire by rocks or thick mud that find their way between the tire and fender.

Aluminum non-clogging rims are by Akront; they're not the heftiest around, but certainly strong enough for trials work and definitely light in weight. Tiny-gauge spokes connect up to a pair of beautiful Dural alloy hubs that are perhaps lighter than *any* and quite strong. The rear hub is new this year and is essentially the same diameter as the front unit.

Though braking force is adequate for trials sections and low-speed applications, it leaves something to be desired when traversing ground between sections at higher speeds. Again, their resistance to dousings in deep water is negligible, and recovery time to dry is far too long. We've found after-market brake linings that do a better job in Bultaco hubs, but due to the unusual sprayed-on braking surface inside the hub, care must be used in selection of lining materials. Some linings can destroy the surface, so caution should be used. It's best to deal with a reputable outfit that specializes in brake work and has Bultaco experience. Rear brake actuation is via a rod, while the front uses a cable.

Our Bul came set up with the brake on the left and shift on the right, which we prefer; but the owner can have it either way when ordering the machine. This is one of the bikes still imported with a right-side shifter.

Small-diameter axles are found front and rear, in the interest of weight-saving, yet the importance of strength is not overlooked. The front axle will be most appreciated by regular trials competitors because it is recessed into the fork assemblies on both sides. The left side of the axle is slotted for a large screwdriver, and a 15mm socket will loosen the axle nut on the right. The recessed ends can't be snagged on obstacles and cost the rider points, nor can the ends of the axle itself be damaged.

At the rear, snail-type chain adjusters are fitted, as is true of other Bultaco models. It's a system we find easy to use, strong and hassle-free. Along with these, a chain tensioner is standard equipment, a must on any type of trials machine. But we'd like to see Bultaco make the tensioner block reversible so that when one side wore out, a simple switch could be made to the fresh side, eliminating the need for a trip to the dealer, who never seems to stock the rubber blocks anyway.

Suspension has been given a going over for 1976. The Betor trials shocks are new and now mounted farther forward, increasing rear wheel travel to 4.75 in.; and compression damping has been softened. This means that the shocks will bottom out far more frequently—in any of the three available preload settings—particularly on faster trails. Bultaco says that's the way it should be, to insure full use of all available shock travel in trials sections. That may very well be, and we like the way the units work in slow sections, but the pronounced "thud" when the shocks bottom along the trail is annoying and a bit disconcerting to the rider. A switch from the standard 55-lb. springs to the optional 70-lb. Miura springs would help.

The shafts on the Betors are also prone to bending without too much provocation, which is why, were the machine our own, we'd make a switch to the special gas/oil Girlings now so popular with many trials riders. But the stock shocks do attach to the machine with lightweight screws instead of the customary bolts; every little ounce pared helps.

Up front we have little to complain about. The Betor forks with the offset front axle work superbly and are no trouble to maintain. The leading axle setup makes for far less steering inertia. But right off we switched to accessory fork springs (that don't sack out nearly as quickly as the stockers), and 10W Bel-Ray fork oil. Bultaco has also improved fork seals;

ours didn't leak a drop. Based on the experience two of our staff members have had in rugged two-day endurance trials, the standard seals are good for about 1000 miles of rough going before leakage begins. Fork travel is 7.0.

Triple clamps are polished and forged of aluminum; Allen bolts clamp the fork tubes in place. Adjustability up and down is possible with the fork tubes, since the handlebar mounts are swept rearward and don't interfere with the top of the tubes. Handlebars fit our staff riders quite well, but we suspect that many owners will switch to the popular Renthal aluminum models that are available in many different bends. Clutch and brake levers are malleable alloy; rubber grit covers protect the hinge point from assorted mung. We don't much care for the grips, but we'll leave that decision up to you.

The heart of the package is, of course, the super-smooth two-stroke Single, displacing 326cc. By varying deep engine components, Bultaco puts its basic engine to work in an extraordinary number of applications, from street to observed trials and just about everything in between.

The crankcase splits vertically, housing a large, three-piece crank assembly that turns in two ball-bearing main bearings. Primary drive is via a single-row chain to the multi-disc clutch. Most Bultacos use a double-row-chain primary, but the larger flywheel assembly required for the T's low-speed grunt and smoothness takes up enough space inside the primary case to eliminate the double-thickness chain. The transmission and primary oil reservoirs each require different amounts and viscosities of oil that can be easily changed in a short amount of time.

Porting is basically as before, with an intake, two main transfer ports, two "finger" transfer ports and a lonely exhaust port. The aluminum piston holds two rings.

Engine mounting is provided by two 19mm bolts at the rear where they connect to a new sandcast aluminum mounting bracket that fits to the swinging arm pivot bolt. This replaces the rear backbone tube that once served as a rear engine>

mounting point. Use of a longer swinging arm is thus allowed, weight is saved, and available space between the carburetor and rear fender is increased.

Up front, a single 19mm bolt holds the engine to the downtube of the frame. For additional support a headstay bracket is used to connect two of the cylinder head bolts to the frame's toptube assembly. Bultaco recommends that a cracked headstay bracket be replaced as soon as possible to prevent additional stress that might fatigue and crack other mounting brackets.

Cylinder and head finning are ample, though many of the fins on the head are cut away to make room for the swept-up exhaust pipe that curls over the top of the engine. The two-piece exhaust system is new and weighs just under 10 pounds. It is secured with three junction springs to the exhaust manifold, a 13mm bolt and bracket to the upper rear engine mount bolt and at the rear by a 10mm bolt. The two joined pipes are held together with yet another spring. And, surprise of surprises, the rear spark arrester portion is actually stamped, "Spark Arrester," which is bound to make somebody happy if they are checking for legality. The new system looks to be very similar to last year's Alpina setup; it's certainly just as quiet. . .wonderfully quiet, we think.

There's a new carburetor this year to replace the former model's 27mm Amal. The rubber-mounted 28mm Bing does its part in adding to the engine's smooth power impulses; it's the smoothest of any trialer we've ever sampled or owned. The carb has an enrichening lever so there's no need for messy tickling either. It's the first Bing we've seen with one. Our test machine was virtually a one-kick starter nearly every time; never more than two kicks were ever required, a trait made to seem even more attractive by the lack of primary kickstarting and an awkward-to-use left-side kicklever.

Once underway there are many features of the Bul that make themselves quickly known. The machine seems exceptionally well-balanced and fell right into place under our test riders. All of us were duly impressed by the throttle response and broad power available. First gear is extremely low and seldom used; most sections are second and third-gear propositions. Fourth is a handy hillclimbing gear, and fifth, of course, is for point-to-point cruising. Shifting is typically Bultaco—that is, there's a fairly long throw of the shift lever with a very positive engagement. The clutch worked in fine fashion and held well in circumstances where slipping was required. Clutch pull is on the light side, as well.

Steering geometry is largely a matter of preference, as are bars and grips and helmet styles. Certain riders just can't seem to feel comfortable or get along well on particular brands or types of motorcycles. So there are no doubt those out there who don't like the steering characteristics of a Bul trialer, just as there are those who wouldn't have anything but. We *will* say that this is definitely a bike on which the very best of trials riders can do business. . .and it can handle beginners just as comfortably.

Though the Bultaco has had the large-bore trials market all to itself the past few years, that will be changing by the time you read this. Montesa is due on these shores any day with its new 310 Cota, and Ossa will follow closely on its heels with

SUSPENSION DYNO TEST

FRONT FORKS

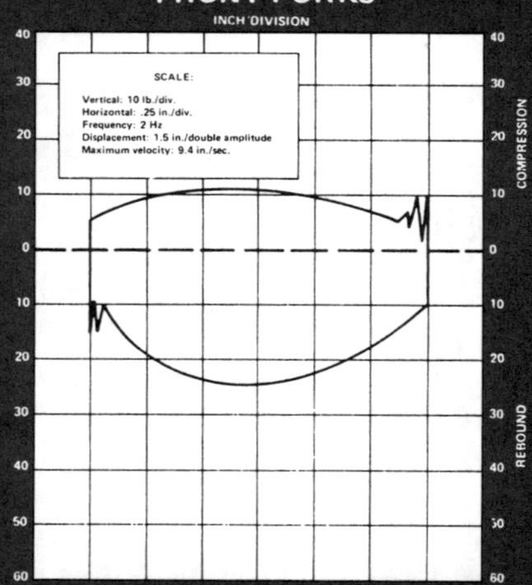

Description: Betor fork with HD 315 oil
Fork travel, in.: 7.0
Engagement, in.: 3.9
Spring rate, lb./in.: 15
Compression damping force, lb.: 11
Rebound damping force, lb.: 24
Static seal friction, lb.: 17

Remarks: Compression damping is good. Rebound damping is heavy in relation to the spring. This is necessary for trials riding because of the slow speed involved and because it is essential to prevent the machine from reacting quickly to major terrain obstacles. The only change we would recommend is fork seals. A set of Universal seals will cut seal friction in half. This will allow the forks to respond more readily to the terrain.

REAR SHOCK

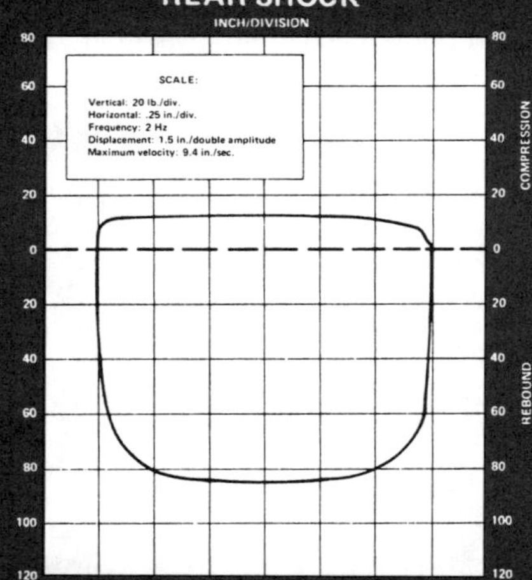

Description: Betor trials shock
Shock travel, in.: 3.75
Wheel travel, in,: 4.75
Spring rate, lb./in.: 47/58 progressive
Compression damping force, lb.: 12
Rebound damping force, lb.: 85

Remarks: These shocks were designed for trials use only. They are extremely light and fragile, especially when a side load is placed on them. Like the front forks, compression damping is light; rebound damping is extremely heavy for the spring. We suggest no alterations. Optional springs are available.

Tests performed at Number One Products

the big-bore MAR. How they will fare against this beauty is pure conjecture at this point, but they'll have their work cut out for them, that's a bet.

Bultaco is still producing the 250 version of this machine at $200 less than the 350's $1498 price tag, and a scaled-down 125cc T will be along soon at $998. Coming from a company that has remained at the top of the trials heap for several years and survived the Japanese trials onslaught without much trouble, we can't help but think this most recent model will hold its own. With the basics down absolutely pat and the infinite details attended to, the beautifully styled '76 T had us all saying, "How are they ever going to top this?"

SHERPA T 350

SPECIFICATIONS

List price	$1498
Suspension, front	telescopic fork
Suspension, rear	swinging arm
Tire, front	2.75-21 Pirelli
Tire, rear	4.00-18 Pirelli
Engine, type	two-stroke, piston-port Single
Bore x stroke, in., mm	3.27 x 2.36; 83.2 x 60
Piston displacement, cu. in., cc	19.9; 326
Compression ratio	9.0:1 (uncorrected)
Claimed bhp @ rpm	18.5 @ 6000
Claimed torque @ rpm lb.-ft.	18.8 @ 4000
Piston speed @ rpm ft./min.	2360 @ 6000
Carburetion	Femsa flywheel magneto
Oil system	pre-mix
Oil capacity, pt.	oil in fuel
Fuel capacity, U.S. gal.	1.7
Recommended fuel	premium
Starting system	kick, folding crank
Air filtration	oil-wetted foam

POWER TRANSMISSION

Clutch	wet, multi-plate
Primary drive	single-row chain
Final drive	single-row chain
Gear ratios, overall:1	
5th	9.9
4th	13.7
3rd	22.5
2nd	29.0
1st	37.7

DIMENSIONS

Wheelbase, in.	52.5
Seat height, in.	31.5
Seat width, in.	7.0
Handlebar width, in.	32.5
Footpeg height, in.	14.5
Ground clearance, in.	12.5
Front fork rake angle, degrees	27
Trail, in.	3.0
Curb weight (w/half-tank fuel), lb.	200
Weight bias, front/rear, percent	44/56

PARTS PRICING

Warranty	60 days
Piston assy.	$39.66
Rings, ea.	5.90
Rear Shocks (each)	29.76
Wheel Rims (bare each)	
front	30.60
rear	39.80
Drive Chain (standard)	14.34
Front Fender	21.18
Rear Fender	22.65
Clutch & Brake Levers (each)	6.68
Clutch Cable	5.96
Throttle Cable	5.30
Brake Cables	5.96
Ignition Parts	
Coil	17.25
Points	4.80
Magneto Assembly	114.00
Air Filter Element	3.96
Rear Tire (standard)	N.A.

Bultaco's "Baby" Frontera: The bike's unforgettable and so is the price! But it's still the... TOP

There's only one thing better than a Bultaco 360 Frontera, and that's a 250 Frontera, says KEL WEARNE. With its competition background and ease of handling and maintenance, it's a perfect bike for ISDT and enduro-type events, where it brings out the best in all riders!

DURING THE past six months a change has come over the concept of enduro riding. It began with the 1975 ACCA conference's approval of running special qualifying events to choose Australian riders capable of representing the country in the greatest motorcycle test of them all, the International Six Days Trial.

This is linked with Australia having at long last joined the FIM (the world motorcycle sport authority) as an independent member nation. Now we're faced with enthusiastic interest in endurance riding — but confusion

TRAIL 250

over ISDT-type events and "enduros" (often thought to be the same, but they are definitely not).

Recently some enterprising ACU clubs from NSW and Victoria got together to try to standardise the running of the ISDT-type competition. These clubs happened to be the ones noted for running the better events and hopefully, with the agreement to follow the basic FIM Two Day Reliability Trial regulations, things will be even better all round. And with the series of qualifiers starting with the Newry trial in Victoria last November, interest for this year is sure to skyrocket — except that the qualifiers will only give us the best riders and will not give us a team. The ACCA has decided that 1977 will be the first year there will be funds budgeted for ISDT activity.

Well anyway, besides having a few two-day events run under FIM rules with gold, silver, bronze and manufacturers' awards to get the feel of what two eight-hour riding days are like, there's been an upsurge in endurance events (such as the Canberra, Wagga, Mittagong and Puckapunyal Army base six hour events and the Winston Stokes

TOP TRAIL 250

Although it's obvious the 250 is less tractable than "Big Brother" it's also far less tiring to ride.

Hare and Hound endurance ride at Singleton). The scene for the two-day trial and enduro buff has been really good in the last six months. (Readers interested in this scene should check BIKE BOOK No 5 — Ed).

So what has all this got to do with a test? Well, just by looking at the top finishers and the number of consistent entries, we find some interesting facts: although there are far more Japanese trail machines sold around the eastern States, guess what sort of machines dominate the gold medals, the club and manufacturers' teams awards and the top places in the endurance events? Right, the European ones, mainly Bultaco, Ossa and Montesa. You don't see any Suzuki or Kawasaki bikes around, and very few Yamahas win. Honda has a fair share with super-keen riders who know reliability will get them home (as finishers, not as winners).

During this last six months we have managed to keep riding the 360 Bultaco which was tested in TWO WHEELS, October, 1975. But the scene changed again recently when Bultaco followed the 360 Frontera up with a 250 Frontera — right in amongst the toughest class of the lot! The 250 enduro class has the Montesa gunning for top berth against the latest Ossa and its impressive heritage of wins.

The big 360 Frontera is superb, but there's a saying that the largest machine one needs for trail and enduro work is a 250 and frankly, we agree. The 250 Frontera will do everything the 360 does and make things a little easier on the rider at the same time. We found the job can be far easier with a top 250 — and the Frontera 250 is definitely among the top two in this most competitive of enduro and trail bike classes.

Similar in dimensions to the 360, the 250 Frontera weighs in at the same 105 kg. It gets long travel Bultaco specification Betor front forks and Telesco units at the rear exactly the way they come on the Mark 8 Pursang, forward mounted along the swinging arm but still vertical.

But all of these things cost more. The latest line of Spanish enduro machines are very much tricked up and competition oriented; not "fun" trailbikes as may have been the case prior to the huge demand for absolute, no-compromise, enduro bikes. The Bultaco factory's answer to this demand is the new Frontera models.

The 250 has a chrome moly frame (it costs) and uses the very latest motocross wheels. These have ultra-lightweight hubs made of Dural alloy, with a thin chrome brake friction area in the hub, stainless steel spokes with aluminium nipples and shoulderless alloy Akront rims. Weight saving is paramount — even down to self-tapping screws rather than a security bolt to hold the tyre in place.

These work well enough for the standard Pirelli tyres, but take a hint; when changing tyres, say to the excellent Metzelers or such brutes as Trelleborgs (not recommended for trail work), a security bolt for the rear is a must. This stabilises the rear tyre and it never moves on the rim. Both the abovementioned tyres do move. The Pirelli tyres aren't all that good, especially on hard, dry going. The reason we recommend Metzelers is simple. The Six Day pattern must rate among the best two enduro-reliability trail tyres available (the other is the Barum SD pattern). It lasts longer than most tyres and provides stable traction under all conditions. And it is tough. In one enduro a 4.50 x 18 Metzeler punctured early in a section and was ridden hard for 23 km over newly formed track consisting of loose sharp rocks to the next checkpoint. In a futile effort to keep within distance of a gold or silver medal (having lost 12 minutes on the section already) Finilec was used to inflate the tyre to about 30 psi. No, it didn't go down immediately, but lasted the rest of the day and stayed up until later in the week when a new tyre was fitted!

But back to the light Frontera 250 wheels. The brake backing plates front and rear are light alloy. The brakes work exceptionally well in the dry but need steady drying out after running through deep water.

With the Frontera range Bultaco started selling its machines with the gear lever on the left hand side, in line with the Japanese. The lever is made of soft alloy and has contributed to missed gears on some machines — it also bends easily in a spill, but this is quite a bit more acceptable than damaging the gear selector forks mechanism.

For those who want the original setup a few hours' work will bring things back to the way Bultaco originally designed them, with the right side gearchange. Even without going to this bother, the 250 Frontera still has the excellent Bultaco litheness and grace over all types of nasty terrain. You don't realise the ability of the Frontera until you see the antics and problems of less able trail bikes. For most riders the 250 does things neatly all the time . . . without the somewhat quicker reactions of its big brother. In the time the Fronteras have been here they have figured in many first placings in enduros and gold medals in two day trials.

Many owners seem to make personal changes. These range from leather or vinyl tank covers to maintain appearance and also to serve as protection in case of a crash among the rocks, to tricky bits like chain tensioners under the swinging arm mounting bracket. It should be there from the manufacturer, especially with the chain whip inherent with the long travel rear suspension. And a side

The 250 Frontera is ordered by many owners with the gear lever changed back from the left side to the right. This 250 is one such machine. Idea is not to use the long flexible aluminium lever which fits under the chain on the left side . . . right side is a short, straight one offering more positive changes.

Right:
Long travel suspension makes the Frontera 250 the easiest enduro machine to ride over really rough ground, particularly rocks, whoop-de-doos and downhill sections. Traction from the 250 is superb.

stand would have been oh-so-nice in many spots. On the current shipment and all to come a sidestand as well as the centre stand is fitted.

Bultaco has stopped sending models with fibreglass mudguards which broke in the first fall or fractured with vibration. The smart ones just changed to Preston Petty Mudders. The next shipment that came along, including the 250 machines, had beautiful stainless steel polished guards which look fine and last, so long as the owner doesn't bend them up against a tree. Very spiffy but it'd be a good idea to take them off, add the Mudders and have a beautiful set of SS guards to use when it comes to selling the bike later on.

Other little things to protect the appearance include using clear Fabulon on the tank. This protects the transfers and the fibreglass tank from scratching. The 250 doesn't get the Finilec tyre pack which has for so long been part of the Bultaco trailbike package. Still, they are easily bought if you want them.

The heavy fibreglass bashplate really takes some pounding. We smashed it time and again on rocks, trees and hard banks without any problems at all. This seems about right for the Frontera line as it has become noted for finishing enduros. No one criticises their frailty.

The 250 is certainly less tractable than the 360, but this is not necessarily a criticism. What it means is simply that a rider can use the throttle with confidence knowing he won't get bitten. The 250 can be ridden harder throughout a day without the rider becoming quite so tired as on the 360. This shows up particularly in rocky terrain where the bike is a flyer. It is brought back to earth only in mud, where some other 250s seem to have an easier time (purely a personal finding from observations at various enduro and two day events).

Starting, long a bug bear on non-Japanese machines, is easy — the same as the Frontera 360: Switch the petrol on, push the mixture enriching lever down, lean the bike over a little and pop the kick start lever back and forth on compression a few times until you hear the fuel take up. This priming will do wonders for the cold start test on two day events.

The enriching lever on the new Amals is ideal for trail use as well — it's not a full choke like Mikunis. When the bike is taken into deep water this lever is handy: Remember, most machines almost stall in deep water as they cool rapidly. Metal changes shape, and no matter how good your waterproofing, the gaskets and sealing rarely manage to keep up with the sudden shock of a hot engine being enveloped by cold water. The Frontera can be ridden through with the lever depressed and then ridden off hard without the bike stopping. In the same way, if the bike feels really hot after walking it up hills in hot conditions (you know, where things get rattly and very metallic), a quick use of the choke cuts a lot of the air off, enrichening the mixture and cooling things off a little. The Frontera will run like that.

If there is one thing a serious enduro-reliability trial rider learns quickly it is maintenance. A few failures bring home the lesson. With the new breed of machines from the lightweight motocross heritage, an owner can expect plenty of servicing and maintenance between events. By doing this he is assured of dependable, reliable riding and maybe lots of medals all round. It's that sort of bike — it brings out the best in all riders.

Enthusiasm for ISDT competition is one thing — winning or placing in the event is another matter entirely. You'll see what we mean on the next page — European correspondent JIM GREENING with the behind-the-scenes story on where the gamesmanship finishes and the real competition begins!

SPECIFICATIONS

MAKE . . . . **BULTACO**
MODEL . . . . **FRONTERA 250**
PRICE **$1395**
WARRANTY . . . . **Three months/3000 km**

ENGINE: Single-cylinder, air-cooled, piston port, two-stroke, alloy head and barrel with pressed-in steel liner, using pressed, built-up crankshaft supported on two main ball bearings, caged roller bearing for big end and needle roller bearing for small end.

Bore x stroke . . . . 72 x 60 mm
Capacity . . . . 244 cm^3
Compression ratio 12.0:1
Claimed power . . . . . 22.4 kW (31 hp) at 8000 rpm
Claimed torque . . . 29.9 Nm (22.1 lb/ft) at 6500 rpm
Lubrication . . . . Petroil-mixture, 25:1
Carburetion . . . . 1 x 36 mm centre float Amal
Air filter . . . . Oiled foam
Ignition . . . . Flywheel magneto with contact breaker points
Starting system . . . . Non-primary kick

TRANSMISSION: Duplex primary chain to constant mesh five-speed gearbox through 13-plate metal clutch. Shift pattern is one down and four up with left side change standard but shaft extends on right side for option. Final drive by 5/8 in. Joresa chain.

RATIOS (Overall :1):
1st 27.48
2nd 17.19
3rd 12.16
4th 9.26
5th 7.6
Primary reduction 2.37
Secondary reduction 3.23

FRAME: Chrome moly welded tubular, single downtube.

SUSPENSION: Front: Betor long travel telescopic forks, internal coil spring, hydraulic two-way oil-damped. 190 mm (7.5 in.) travel. 30 deg rake, 152 mm (6 in.) trail.
Rear: Swinging arm, Telesco forward-mounted spring/damper units with 140 lb springs, 165 mm (6.5 in.) movement.

WHEELS, TYRES AND BRAKES:
Front: 3.00 x 21 in. Akront shoulderless alloy rim. Pirelli fully knobby tyre, 140 mm x 30 mm single leading shoe drum brake.
Rear: 4.00 x 18 in. Akront shoulderless alloy rim, Pirelli full knobby tyre 140 mm x 30 mm single leading shoe drum brake.

ELECTRICS:
Source . . . . Flywheel magneto
Lighting . . . . 6v 40/35 W headlight 18/5 W stop/tail light

CAPACITIES:
Fuel tank . . . . 12 litres (2.6 g)
Primary chain/clutch . . . . 300 cc (SAE 30)
Transmission . . . . 600 cc

DIMENSIONS:
Dry weight . . . . 105 kg (231 lb)
Wheelbase . . . . 1384 mm (54.7 in.)
Overall length . . . . 2083 mm (82 in.)
Footpeg width . . . . 446 mm (17.7 in.)
Overall height . . . . 1118 mm (44 in.)
Seat height . . . . 890 mm (35 in.)
Footpeg height . . . . 368 mm (14.5 in.)
Ground clearance . . . . 260 mm (10.2 in.)

PERFORMANCE

HOW DO THE FINER POINTS SCORE (Out of 10)

Instrumentation 7
Comfort 8
Tool kit 7
Quality of finish 8
Vibration 9
Lighting 7
Positioning of controls 9

TEST BIKE SUPPLIED BY: Bert Flood Motorcycles, 1111 Whitehorse Rd, Box Hill, Vic.

STREET VERSUS TRAIL

Sit Down Riding With Fingertip Control

BY DAVE EKINS

It's general knowledge among my circle of friends that I'm some sort of Bultaco freak. I have been competing on them almost exclusively for the past five years. I had one of the very first Alpinas, engine number 00002, and two others since then. I fell out of love with the Alps when they put them in a frame of their own. That's when I tried a 250 Pursang converted for serious enduro competition. Now that was a bike. First model of the long travel Buls, it worked out to be the best cross-country mount I have ever thrown a leg over. I kept the close-ratio gearset and just slipped the all-metal clutch when the going got sloppy and slow. It was a great cow-trailer although a bit costly for the average guy to undertake.

During my affair with Bul I herded one of the first Fronteras and had the next-to-last Matador, the Mark7. Matadors always intrigued me because they were intended for the serious ISDT rider. "A bike you could win Gold Medals on," they used to say. It had great tractable power for a 250 of that era, excellent handling and suspension. But its nicest feature was the fully enclosed rear chain. I had ridden a pair of back-to-back two-day ISDT qualifiers with this bike and never did adjust the chain. Didn't do too badly, got a Silver and a Gold for my effort.

The reason for all this first person dialog is to qualify me to talk about Bultaco's new era Matador. It is no longer a motorcycle to ride an ISDT with. They have built a bike for the lay-back woods-and-trail rider who wants to leave his front door, ride city streets, and take a journey off-road all on the same machine. But, if you want to get serious and compete in enduros, the bike is capable with just a few modifications. In this sense it is far ahead of its Japanese-made street-legal enduro counterparts. Because the Matador, even with its tack-on goodies to meet highway regulations, is still of the fine Bultaco heritage that sets them apart from all other motorcycles.

Until now Matadors have been 250cc displacement machines; the MK9 is a 350. It has the 348 Alpina engine and gearset, and the offset Alp carburetor and complete exhaust system, with two mufflers and boomerang connector. Alpinas are essentially Sherpa T motors with Matador even-step wide-ratio gearsets.

The Matador has Alpina front axle forks and the Frontera/Alpina front wheel. Rear shocks are the new Betor gas pressurized types similar to what is found on the Pursangs. Front wheel is one of the lightest available, even with a speedo drive. It has Bultaco's exclusive hard chrome plated wear surface for the brake lining to rub on. Other bikes have the less

New carb for Bul. Bing instrument replaces Spanish Amals, left side shift ignores new boss in casting.

Removal of fiberglass cover reveals Japanese-made battery sans plastic overflow tube which someone left out.

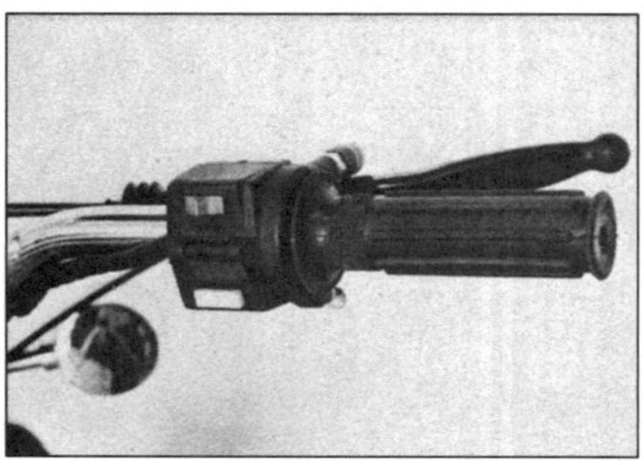

Boomerang muffler is standard for Sherpa T and Alpinas; it does the job for torque-tuned engines.

Throttle, brake lever, and switch assembly are all built into one flat black finished aluminum casting.

Front brake cable is not covering the 10th-mile readout window, there isn't one. Switch assembly in left side has two positions, on and off. Key also fits fork lock tumbler located on frame behind headlight.

costly and heavier cast-in steel liners.

The rear wheel, however, is not as lightweight. It is the old Matador quick-change unit complete with rubber cushions between the driving hub and wheel. It has the steel liner. Add this to the chain enclosure casting and rubber tubes the chain passes through and the thing has gained a few pounds. The bike is a left side shift, right side brake model and because of this they can use a brake rod instead of the cable formerly used. My initial ride on the bike told me the rear brake was like stepping on a brick with no slowing results. Being an old Bultaco freak, I just disassembled the back wheel and sanded the brake shoes. Now it has velvet touch brakes, well, almost.

Electrics seem to be the result of several international trade agreements. A Spanish Femsa flywheel magneto/generator churns out the juice. A Japanese Yuausa 6-volt battery stores it. Head and taillights are Lucas of England, and the turning indicators are Goneli of Spain. Handlebar control switches are in combination with the front brake and clutch levers; they are Lucas. These are the ones with the easy-to-flip flags that can get you in the tender part of your hand. Hi-Lo beam flag is on your left side with the light switch off, park, and on, a sliding switch above. A button under the flag operates the hooter. Turning light switch is flag operated on the right-hand side with a kill button above and starter below. Now if you visualize what I've just said then think about all those other bikes, you will realize they have switched hands on us. Taking this one step further and you realize the switch is British and the British drive on the other side of the road.

Staying with this standardization theme Bultaco has put the shift lever on the left side, a place not to F.X. Bulto's liking. He feels the clutch should be opposite the shift and the brake opposite the throttle; just as you swing your left arm forward when you move your right foot in the same direction. He says this is Nature's way and the interchanging of shift and brake levers is against Nature's original design.

All recent Bultacos changed their shift to the left side simply bypassing the shifter shaft through the left-hand case. The Matador has a fully enclosed rear chain thereby presenting problems with doing it this way. So they ran a shaft through the frame, attached a brake lever to one end and a short arm to the other. There is a similar arm clamped to the shifter shaft coming out of the engine on the right-hand side. A short rod connects these two shafts. A simple and effective solution to a problem.

In an effort to keep the price down, mild steel tube is used for frame construction. With only weight being the disadvantage over chrome moly the advantages are less probability of failure, it can be welded without using a special inert gas process and the price is right. You sacrifice weight, but that isn't a problem with the intended use of this bike. The gas tank is steel as are the fenders. The only plastic items we could find are the hand laid up side covers, air filter container and seat base.

This bike is an incredible first kick starter. A bypass valve on the Bing simplifies cold starting. Only problem is for those not used to starting a bike whose kick starter pedal bottoms on the footrest and you have to stand off to one side in order to operate the lever. But then I know how.

The bike is exceptionally quiet and has absolute flat torque. Power comes in at idle and smoothly runs up the scale. The shifter takes a little more effort than a Japanese bike and the clutch pulls a little harder. Brisk acceleration making quick shifts gets you into fifth gear while going only 40 mph. It is completely undone at 60. The thing is a tractor with tractor gearing, not that they wanted it that way, it is just the way it turned out using the existing 250 Matador pieces. Suspension is the best of any street-dirt bike enduro I've ever ridden. Betor forks soak up everything from tar strips across pavement to chuck holes on the Macadam. These new rear shocks are just as worthy to the task. The bike is not one you would motocross or special test. It is a sit down and ride it model. And the comfortable seat keeps you right there. Steering is fingertip light, too light. I was able to wash the front tire, even on pavement just by oversteering it. If you pack another, it is going to be that much worse.

The passenger has a pretty good lot while perched on the back of the Matador. He has a grab rail around the seat; and footrests in the proper location but fastened to the wrong thing. Passenger footrests should not be mounted on the swing arm. It's almost acceptable on paved roads. But he will never keep his feet on them off-road. Rear shock spring setting is fine in the soft position for solo riding and there are adjustments for more stiffness to accommodate another person. The Alp engine does not produce sudden hp so dumping the rider off the back would be difficult. In fact, trying to get wheelies out of the thing is a chore. Like I said, the Matador is to sit back and ride. And for those persons there are only two problems that seemed distasteful to me. One is the engine has to run on pre-mix which isn't a problem in Spain where you just pull right up to the 20-to-1 pump and say, "Fill 'er up." The other is the amount of wind noise generated in the air filter area. With its enclosed chain and double muffling it should be the quietest machine running. It isn't. There is so much noise coming out of the air box that you need ear muffs when riding. It seems to me the people at Bultaco fell down right here. So there you have it; a lay-back enduro with fingertip control.

"...The Bike Is Exceptionally Quiet And Has Absolute Flat Torque..."

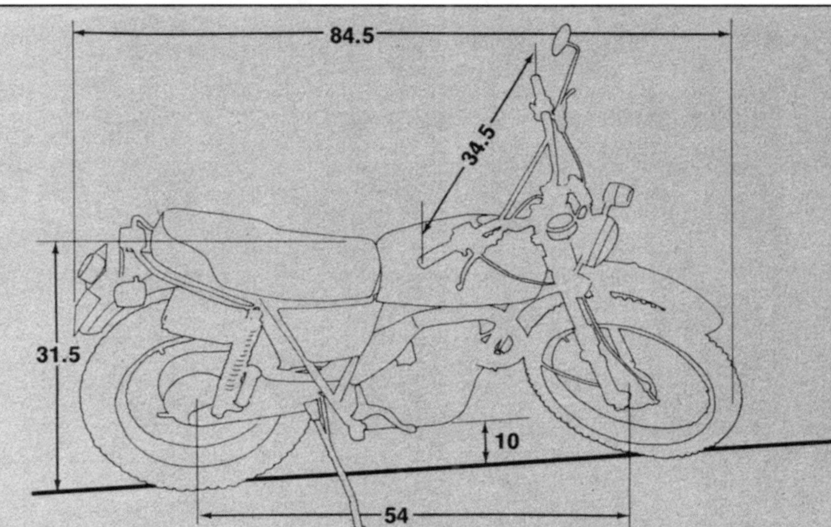

BultacoMatador

TEST BIKE: BULTACO MATADOR MK9
Base price.............................$1525

ENGINE
Type...........*Two-stroke, piston port*
Displacement.........................*348cc*
Bore x stroke..................*83x64mm*
Compression ratio.....................*10:1*
Lubrication system.............*Oil mist*
Carburetion..................*32mm Bing*
Air filter.........................*Oiled foam*
Ignition system...*Flywheel magneto*
Starting....................................*Kick*

DRIVETRAIN
Primary/ratio..........................*2.37:1*
Gear Ratios.....*(Internal) 1st .287:1; 2nd .442:1; 3rd .625:1; 4th .821:1; 5th 1:1*
Clutch..............*All steel multi-plate*
Final drive/ratio.....................*3.23:1*
Sprocket teeth......................*13/42*

CHASSIS & SUSPENSION
Frame..............................*Mild steel single cradle*
Trail..*5.5"*
Suspension,
Front.......................*Telescopic 7.2"*
Rear.........................*Swing arm 5.0"*
Brakes,
Front....................*125mm dia. SLS*
Rear....................*140mm dia. SLS*
Tires,
Front..........................*2.75x21 Trials*
Rear..........................*4.00x18 Trials*
Rim locks,
Front/rear.....................*Rim screws*

WEIGHTS & CAPACITIES
Weight, wet, unladen...*269 pounds*
Fuel capacity.................*2.6 gallons*

Bultaco Frontera

BY MARTIN CHRISTIE

Bultaco have an advertising campaign in American magazines at the moment. It features a half-page picture of a sewing machine, upon which the Bultaco symbol has been superimposed. The caption says 'No Way'. And the blurb goes on to say that the Spanish factory just builds motorcycles and nothing else. If it didn't, it wouldn't be able to build them so well, and so on....

It's not an unfamiliar line of patter. And for all that there is something behind the reasoning. But its one of those somethings that a lot of people, and in particularly advertising executives, would have difficulty defining.

It defies the bland phrases and the consumer-orientated compromise. It is simple, raw motorcycling. No frills, no fanfares, no 'fabulous' offers.

It's a gut reaction that tells you, begs you, when all common sense, and on balance weighing up of the pros and cons, would say no. Bultaco's particular sewing machine campaign is a rather shameless dig at a factory not a million miles from the town of Huskvarna in central Sweden, which also builds a lot of other things including kitchen units and chain saws.

But it's a poor choice of victim. For Bultaco have too much in common with several relatively small European factories to wave the finger of sarcasm. They'd be better to turn their jibe further east and use a portable electricity generator, a Jeep or/and orientated shipyard for comparison, and contrast.

Compared to the sleek, smoothly finished, chromium plated products from Japan, a real European bike like Bultaco has a ragged, cobbly look that could draw unfavourable criticism. Those little details which were not quite right. One or two points that could be improved with a bit of attention.

But to single out minor detail is unkind and almost unfair to the understanding of the whole concept. It is an enthusiast's machine. That in itself is something of a euphemism. It means you're prepared to put up with a series of niggling faults — all to live the sheer experience of a totally unique machine.

Bultaco's Frontera is such a machine. A bike that just defies any attempt to do a serious or objective test.

One of few others with this temperament is probably Laverda's Jota. A totally uncompromising motorcycle to the point where it becomes agonising ecstacy just to ride it. To have enjoyed it and then condemn it because the switches were tatty and the seat too hard, is to have drunk champagne and written it off as fizzy cider.

Bultaco's Frontera makes no concessions. It is an out and out racing motorcycle. Like all the other machines imported into this country by Bultaco UK it is intended primarily for competition.

Hand-built in Barcelona at a factory begun by bike-mad brothers from an old farmhouse. From those early days after the war it has grown to cover a larger production, and take in a wider range of models — particularly small capacity commuters for the protected home market. But the first love, and the overwhelming passion is cross-country competition.

And that passion is reflected in the flair and the panache of the Frontera.

Unlike the more modest, and rather overweight Matador, intended for the trail or more sedate enduro, the Frontera is a purpose-built ISDT bike. Basically it's one of their Pursang scramblers equipped with minimum items for road legality, and slightly detuned for overall regularity.

The frame is the same very light silver grey sprayed chrome moly tube, with a single downstem widening to a double loop cradle under the engine. Staggered just aft of the main spine tube under the seat, the rear sub frame supports twin gas Telesco units with dual rate springs in a semi upright position. The only addition to the tubing is a couple of hand grabs tacked on behind the seat for manhandling moments.

The bottom mounting for the rear units is mid way along the tubular swinging arm, with additional plates welded on underneath both arms for extra strength. With around eight inches of movement for the rear wheel allowed for, this makes the seat very high at 35 inches, and there is no easy way to change the angle of the units to

Bultaco Frontera

reduce this as on some other makes.

A single prop stand which folds along the left side of the swing arm is the only permanent support for the bike, although there are two frame lugs on the lower main loop which fit a square section paddock stand for more sturdy security.

The 250cc version of the Frontera has a single side mounted mainstand which can hold the bike upright, but this rather awkward fixing has been dispensed with on the 370, and in any case the simple prop still allows the bike to be tilted over for rear wheel removal.

The latter operation, essential for six day trial work, can be achieved by the simple loosening of the axle and brake, leaving the chain in place.

The other main difference from the two-fifty is the front fork. Instead of the shorter fork, of Bultaco's own manufacture used also on their 125cc moto cross bikes, the 370 uses the much longer ones with forward mounted axle employed on the 250 and 370 Pursangs.

This gives a massive 9¼ inches of movement which is essential for the moto cross track but may be a shade too much for overall time trial work. Certainly Bultaco supported riders using the big Frontera have opted to use the shorter forks in competition, and one of the reasons is the tendency for the longer fork legs to dig in when ploughing through deep ruts.

On tarmac the extra movement, with the first few inches being relatively soft, was something of an embarrassment with the bike steering rather erratically unless under power. Once those inches were taken up, there was no problem. But otherwise the wallowing of the front end and the resulting vagueness of the steering would make cornering nothing if not entertaining.

Of course using the Frontera on the road is something of an embarrassment anyway, it's not meant to be there, it's not really built to be there. You just have to tow it somewhere you can find room to let it loose. Any road sections you have to negotiate in between should be done with care and discretion.

Not for safety's sake — but mainly for fear of antagonising local residents and worse, the law. Because it is noisey.

It has a massive bazooka-style exhaust system that begins at the base of the sub frame and aims itself diagonally up the right side of the bike. If looks were the key, it should do the trick. But it just fails to cope with that fiendish two-stroke crackle. And for continuous use, you'd just better look for understanding neighbours.

As far as effectiveness goes, it would probably be as well to fit a shorter and light pipe of the type fitted to the moto cross bikes. It can't produce much more noise, and it will certainly tuck out of the way better. For despite a massive glass fibre sump cover that extends over the beginning of the exhaust system, there is a danger of clouting the whole affair when clambering over rocks.

The exhaust system itself is readily detachable. Three springs secure it at the head — tricky to get off under pressure, but surprisingly easy to get back on again once you've got the knack.

The only other mountings are a stay which secures the tail of the pipe to the rear frame, and a rubber bung tucked away behind the rear brake pedal. You'll need to know this because the original black coating is wafer thin and wears through to bear metal within a few days use, so you have to get the old engine black out to restore it to pristine condition.

The bike Bultaco UK lent me was absolutely new. They'd just taken it straight out of the showroom, registered it, and taken it down the road to make sure it ran. It had exactly two miles on the clock when they handed it over, which shows either great trust in me, or more likely great confidence in the machine.

Whichever, I was impressed. Over three weeks of varied use, including several days charging round the Welsh mountains, it behaved impeccably. Run on ordinary Shell 2T at 25 to one as recommended it smoked a lot but never once fouled a plug. And every prod at

the forward mounted right-foot kick produced a first time start.

The only difference from the 363cc scrambles engine is a lower compression ratio, and the use of only one plug with the independent coil Femsa flywheel ingnition. From such a fire-breathing engine, such sweet reliability was scarcely credible.

Bultaco said running it in was unnecessary. Just run it well. So, perhaps a little apprehensively, I increased the pressure on the throttle over the first few days. There was no hesitation, and no complaints. The only thing that took some bedding in was the gearbox. The close ratio five speed cluster took a few

days of easing before selection became automatic, though the left side pedal remained rather stiff in operation.

Starting from cold the clutch also needed to be treated with caution to free all the plates before use. On the move, changing with or without the unit, was faultless.

The big 36mm Bing carburettor — used on the Frontera in preference to the Spanish Amal — gave very smooth running characteristics and metered the fuel without vices. But it did drink it up at an alarming rate, inevitable with a large capacity two stroke.

And that leads directly to the only serious problem with the Frontera — the only problem that can't really be dismissed as inevitable with a competition-designed motorcycle. The petrol tank was too small. Bultaco build them in Spain with a glass fibre tank, but of course our current regulations make that illegal on the road. So Bultaco UK had to have a small batch of tanks specially made in alloy.

The tank has an attractive three-dimensional wedge shape, and would be almost large enough as a fuel reservoir were the filler cap not placed too far down the top face. As it slopes away from the securing bolt at the neck, it means there is an air pocket at the highest point of the tank which you can't fill because it would run out of the cap. As it is, the tank will take exactly 1.6 gallons, which means it will average 36 miles on main tank and extend to precisely 42 miles with reserve.

Unless you've got a support crew, this makes long jaunts into the countryside somewhat limited.

There's not a lot of point in riding slowly to conserve fuel, as you might as well ride an ordinary trail bike. It's just one of those things you have to put up with — or get a larger tank.

The direct lighting is controlled by two identical and simple switches on either side of the handlebar. These have a press button for an engine cut out, which was fairly inaudible over the noise of the engine anyway. Two thumb flip switches master the headlight and dip beam of the small 25/25 watt headlamp, mounted on rubber in a number plate nacelle on the fork yolks. Again the lights are street legal-rather than useful items.

The front mudguard was effective as a unit, with extended lips to catch flying dirt, though at the rear, through a slightly imperfect fit, it fouled the front of the exhaust pipe. It wasn't helped by one of the three retaining bolts — two of which also act as lock stops — dropping out from under the bottom yolk.

That was my own fault as I missed it during the essential tightening checks on a new bike over the first few days. Nothing else came loose though, and not a spoke came out of tune.

The only thing that suffered fatal damage was the chain guard, a simple plastic affair, which seems to destroy itself through vibrating on its own mountings as much as being thrashed by the chain. The latter is kept well under

Bultaco Frontera

check by a clever self tensioner.

The Bultaco comes supplied with Pirelli moto cross tyres which give excellent grip under all conditions except tarmac, where, with the power of the engine, rear wheel spin is a regular performance. They are mated onto a yellow label Akront alloy rim on the front and a green label heavy duty on the rear. Both are non valanced and have security bolts.

The single leading shoe drum brakes produced excellent performance all round, even coping with its 90 mph road potential. In common with other manufacturers now Bultaco have the rear pedal Japanese style on the right. But it is still possible to transpose the system and mount the cable operated unit on the left side of the frame. The gear selector shaft still runs the whole way through the back of the crankcases, and is protected on the right side by a simple rubber bung. More traditionally minded folk may prefer the transition.

To talk seriously about riding Bultaco's Frontera though is so much harder than to point out its features, and perhaps its flaws. It does have its limitations, but for the handful of people who are even considering paying a thousand pounds for one day they are irrelevant. It is an extraordinary, and an awesome motorcycle, which will bring most riders to the realisation of their own limitations.

It is a bike that demands more than you can normally give; asks to be ridden harder and faster than you have the skill or the courage to take it.

But because of that it is one of the few motorcycles that has the potential to raise your own expectations — push you to try that bit harder, if only for the self-satisfaction of knowing you can do it.

There is no sense talking about the Frontera alongside an average trail bike, or even a good enduro. It is in a completely different league, and one within which few god fearing mortals can hope to tread.

It left me with a tremendous feeling of my own inadequacy, but with a burning desire to attain new heights.

One day I'd like to be able to ride it properly. Snr Bulto, think of me this Christmas . . .

Specifications

Bultaco 370 Frontera

Engine: 363cc (85 x 64mm) two stroke single. Compression ratio 9:1. 36mm choke Bing Carburettor. Wet foam air filter. Claimed maximum power 34bhp at 7,000rpm. Maximum torque 3.5 at 5,000rpm.

Transmission: Duplex chain primary drive to five speed gear box. Ratios: 0.38, 0.51, 0.67, 0.83, 1.0 to 1. Chain final drive.

Electrics: Femsa flywheel magneto with independent coil. Direct lighting to 25/25 watt headlamp; pilot light, stop light. Spark plug — Champion N3.

Frame: Semi-double cradle chrome moly tube.

Suspension: Bultaco telescopic front fork with fore mounted axle. 9.4in movement. Telesco twin rear gas units with external reservoir.

Wheels: Akront alloy non-valanced type.

Tyres: Pirelli moto cross. 3.00 x 21in front. 4.50 x 18in rear.

Brakes: Single leading shoe drum 140mm x 30mm front and rear.

Fuel capacity: 1.6 Imp. galls including reserve. Petroil measured at 25 to 1.

Wheelbase: 55.3in.

Seat height: 35in.

Ground clearance: 13in.

Dry weight: 240lbs.

Price: £997.00.

Importers: Bultaco (UK) Ltd, Portsmouth Road, Thames Ditton, Surrey. Tel: 01-398 5531.

BULTACO ALPINA 250 AND 350

A Woodsy Pair with Limitations, Quirks and a Stack of Bonuses

Cycle World Enduro Test

When you stop and think about it, most of us ride motorcycles for fun or pleasure and many of us are simply not too serious about riding. The machine doesn't have to be the prettiest or the fastest or the most expensive or have three miles of suspension travel front and rear. All that's required is that it be basically reliable and hassle free and be able to deliver an afternoon's or day's ride into the boonies for the pure enjoyment of it . . . nothing else.

It just so happens that Bultaco builds a motorcycle that slides right into this notch. On one hand it's a special purpose machine and on the other hand, it's not. Let us explain.

Alpinas in various displacement categories have been coming out of Spain for a number of years and for the most part they have been trail machines designed for riding fun. Mind, Bultaco produces a number of other motorcycles for special purposes, namely for motocross, enduro and trials. These are serious motorcycles with few compromises and plenty of specifics in mind. But the Alpinas are really not that way.

Though, yes, one can certainly successfully compete in an enduro on an Alpina, we'd venture to say the majority of Alps sold in this country are used strictly for pleasure, or are used for pleasure most of the time. And when we speak of competing in enduro events, we shouldn't generalize too much. If the event has high speed averages and predominantly wide-open terrain, the Alp won't exactly be in home territory. Throw in lots of trees and rocks or tight trails, though, and the story takes a turn for the better.

In southern California we have to head for the mountains to make the most of the Alpina's design; in the east one could head for the nearest woods and have a ball. The point is, the Alpina isn't really an all-around off-road machine, but if you have the right territory, Bultaco has the right bike.

We're examining two different Alpinas here, different in the sense that one has a bit more displacement and red paint instead of blue. And aside from picky details, what you get with one you get with the other; they're that close. Preference comes into play here, some folks like 250s, others go for a bit more grunt and power, and the 350 has more to offer in those terms.

Differences in the 1976 and 1977 models couldn't be more minimal. Bultaco has changed the paint scheme around and that's it. Bigger changes took place from the 1975 to 1976 versions, where the bike went from a trail motif and leaned back once again to trials, borrowing the chassis and many components from the Sherpa T trialer, which was the original concept when the bike was first introduced.

If you were to stand the latest versions of the Alpina and Sherpa T side by side, you'd be hard pressed to find differences in the basic machine. They share most of the essentials and change a bit when they get to cosmetics. A trials bike needs to be narrow and slim so the rider can move around easily without interference from the bike. That's why trialers have tiny fuel tanks and small seats. But tiny fuel tanks and small seats don't do the job out on the trail. Here you need a comfortable seat and enough fuel to get you back to where you started. So the Alpina comes with all the items needed to turn a trials machine into a perfectly livable trail bike. The seat is large and cushy, suspension is geared toward faster trail speeds rather than slow plonking over obstacles, passenger pegs are included and there is more fuel capacity.

The chrome-moly frame has the same geometry as the Sherpa T model, but the triple clamps are different and they alter steering characteristics. The swinging arm has been altered so the lower shock mounts are further forward, changing the shock position and increasing the rear wheel travel to a usable 6 in. The passenger pegs also mount on the swinging arm, but the bike really isn't designed for two people and the pegs are there for emergencies. As before, chain tension is adjusted by means of eccentric cam adjusters which are fool proof and quick. To smooth power impulses, a chain tensioner is standard equipment.

Because the rear frame loop doesn't extend over the rear fender, Bultaco has fitted a small chrome grab bar which comes in handy when lifting the machine or pushing it, particularly in and out of a truck. The seat itself offers more padding and the slope is changed to help prevent the rider from sliding forward. The 2.24-gallon fuel tank is slim enough to stay out of the rider's way, yet carries enough gasoline to take the bikes 50 to 60 miles. Fuel mileage doesn't seem to vary much between the 250 and 350, perhaps because the larger engine is working easier in comparable situations.

Suspension is identical on both ma->

CYCLE WORLD photos

chines and is manufactured by Betor. Check the dyno chart for the particulars. The forks have ample amounts of travel and respond well at slow speeds, allowing the rider and machine to glide over some rough territory without affecting control. The rear shocks work well in conjunction with the fork action, but at high speeds or really abrupt landings the units will bottom. Riders who plan to ride at higher speeds most of the time might want to look into a higher viscosity of fork oil or stiffer rear springs. But those who spend most of their time plonking or going on the easy side of things will probably feel at home with the suspension in its present form.

Rims are the familiar shoulderless variety made by Akront and are extremely lightweight. They'll survive a good beating in spite of the fact that they look fragile. Once dinged up, we'd probably make a switch to the stronger DIDs and lace them to the spiffy hubs, which are not only tough

Overall finish is typically Bul which is to say, good. Gas cap is easily gripped, rubber lever covers prevent dirt from entering teflon lined cables.

Left side panel is strictly decorative. Both machines feature chain tensioner, left side shifting and the Sherpa T chassis.

250

SPECIFICATIONS	
List price	$1195
Suspension, front	telescopic fork
Suspension, rear	swinging arm
Tire, front	3.00-21
Tire, rear	4.00-18
Engine, type	two-stroke Single
Bore x stroke, in.; mm	2.80 x 2.36, 71 x 60
Piston displacement, cu. in.; cc	14.52, 238
Compression ratio	9:1
Claimed bhp @ rpm	14.1 @ 5500
Claimed torque @ rpm lb./ft.	14.75 @ 4000
Piston speed @ rpm ft./min.	2360 @ 6000
Carburetion	26 mm Amal
Ignition	flywheel magneto
Oil system	premix
Oil capacity, pt.	1.3
Fuel capacity, U.S. gal.	2.24
Recommended fuel	premium
Starting system	kick, folding crank
Air filtration	oil-wetted foam
POWER TRANSMISSION	
Clutch	wet, multi-disc
Primary drive	duplex chain
Final drive	# 520 single-row chain
Gear ratios, overall:1	
5th	7.655
4th	9.313
3rd	12.246
2nd	17.295
1st	26.598
DIMENSIONS	
Wheelbase, in.	50.0
Seat height, in.	33.1
Seat width, in.	6.5
Handlebar width, in.	32.7
Footpeg height, in.	15.7
Ground clearance, in.	12.0
Front fork rake angle, degrees	N.A.
Trail, in.	N.A.
Curb weight(w/half-tank fuel), lb.	224
Weight bias, front/rear,percent	44/56

but about the lightest around. Brakes on both test machines seemed to be an improvement over past models, but they still have problems with water. Recovery time after a dunking seems excessive, but when dry, both front and rear brakes stopped the bikes well and allowed the rider the proper amount of feel during heavy braking.

Neither bike wept fork oil or leaked engine oil, perhaps owing to an improvement in quality control, as both machines seemed to be put together quite nicely. There are still the usual Spanish drawbacks such as aluminum fenders and the lack of primary kick starting, but much of the simplicity is welcomed. Actually, though we complain about the bendable aluminum fenders, with some care they can be kept in good shape and protect the rider well from mud and such. But once again, when it came time for replacement

Alpina 250 PARTS PRICING	
Warranty	60 days
Piston	42.85
(1) Set Rings	11.80
Rear Shock	38.47
Wheel Rim (bare)	
Front	39.47
Rear	40.02
Drive Chain (standard)	19.00
Front Fender	16.72
Rear Fender	32.52
Clutch or Brake Lever	16.88
Clutch Cable	6.00
Throttle Cable	6.00
Brake Cable	6.00
Ignition Parts	
Coil	17.90
Points	4.80
Magneto Assembly	114.00
Air Filter Element	4.50
Rear Tire (standard)	28.00
Headlight Bulb or Sealed Beam	N.A.
Taillight Lens	N.A.
Battery	N.A.

Betor nitrogen gas shocks need heavier springs to prevent bottoming during trail riding. Aluminum fenders will last with care but plastic is superior. Tool box could be designed better.

Alpina 350 PARTS PRICING	
Warranty	60 days
Piston	42.85
(1) Set Rings	11.80
Rear Shock	38.47
Wheel Rim (bare)	
Front	39.47
Rear	40.02
Drive Chain (standard)	19.00
Front Fender	16.72
Rear Fender	32.52
Clutch or Brake Lever	16.88
Clutch Cable	6.00
Throttle Cable	6.00
Brake Cable	6.00
Ignition Parts	
Coil	17.90
Points	4.80
Magneto Assembly	114.00
Air Filter Element	4.50
Rear Tire (standard)	28.00
Headlight Bulb or Sealed Beam	N.A.
Taillight Lens	N.A.
Battery	N.A.

Leading axle Betor forks are similar to the units on the Sherpa T, but have different damping rates. Akront rims and Pirelli tires are usual Bultaco items.

Seat has been slightly redesigned and chrome grab bar added. The bar is a big plus when lifting or moving bike manually.

we'd go to plastic fenders.

The airbox unit is pirated from the Sherpa T and contains a Twin Air foam element accessible from the right side of the machine. Instead of screws to retain the side panels, Dzus fasteners would be a better choice. Same goes for the toolbox mounted on the left rear side of the machine. It just seems odd to have to use a tool to get to the tools and that's the case here.

If you happen to be climbing from Brand X to one of the Alpinas, the Alps might seem very strange indeed. Most of this feeling comes from the high footpeg location, which requires more of a bend at the knee when in the seated position. High mounted pegs do come in handy in rocky terrain or deep ruts, even water crossings. The chances of hanging them on obstacles is slim, and the heavy serrations on top of the pegs keep boots from sliding off readily. They fold up and back on impact.

Looking at efforts taken to tuck things out of harm and rider's way, one can't help but appreciate the exhaust system. It snakes its way out of the exhaust port of the cylinder and curls upward over the two-stroke Single and back behind the frame rails, finally terminating behind the right rear shock. Quiet comes from the muffled center portion and a measure of safety is added because of the rear box portion, which is a spark arrestor. And that means it's legal where spark arrestors are required. The system can be removed easily and can be taken apart for decarbonization. It really is a pleasure *not* to hear this machine and still have ample power delivery despite the increased muffling.

Our two machines came fitted with Pirelli tires, 4.00 x 18 in the rear and 3.00 x 21 up front. The Pirellis are okay, but some riders may want to opt for their own >

350

SPECIFICATIONS	
List price	$1395
Suspension, front	telescopic fork
Suspension, rear	swinging arm
Tire, front	3.00-21
Tire, rear	4.00-18
Engine, type	two-stroke Single
Bore x stroke, in.; mm	3.27 x 2.52, 83.2 x 64
Piston displacement, cu. in.; cc	21.23, 348
Compression ratio	8.5:1
Claimed bhp @ rpm	21.6 @ 6500
Claimed torque @ rpm lb./ft.	20.1 @ 4500
Piston speed @ rpm ft./min.	2940 @ 7000
Carburetion	28 mm Bing
Ignition	flywheel magneto
Oil system	premix
Oil capacity, pt.	1.3
Fuel capacity, U.S. gal.	2.24
Recommended fuel	premium
Starting system	kick, folding crank
Air filtration	oil-wetted foam
POWER TRANSMISSION	
Clutch	wet, multi-disc
Primary drive	duplex chain
Final drive	#520 single-row chain
Gear ratios, overall:1	
5th	7.655
4th	9.313
3rd	12.246
2nd	17.295
1st	26.598
DIMENSIONS	
Wheelbase, in.	51.6
Seat height, in.	32.3
Seat width, in.	6.5
Handlebar width, in.	32.7
Footpeg height, in.	15.0
Ground clearance, in.	11.8
Front fork rake angle, degrees	N.A.
Trail, in.	N.A.
Curb weight (w/half-tank fuel), lb.	224
Weight bias, front/rear, percent	44/56

personal choice of rubber. The standard tires work better if the ground is more damp than dry. The tires are secured to the rims by locking screws rather than rim locks. The screws are pointed and actually screw into the bead of the tire, keeping it from slipping when lower pressures are used. The trouble with the screws is that they are soft aluminum. The Phillips head slots strip easily, making the screws nearly impossible to get out. And without some sort of thread-locking solution on them, they have a tendency to get loose and fall out. But they're less of a hassle when changing tires in spite of their shortcomings.

A speedometer is standard equipment and has a resetable trip odometer to keep track of mileage. The unit is driven from the front wheel, unlike older models which were driven from the rear. Lighting kits are available for those who require them, but the lighting is barely adequate.

The 250- and 350-cc engine units share many components and are unmistakably Bultaco. Primary drive is via a duplex chain with tensioner, and a Femsa flywheel magneto ignition provides six volts of juice. Flywheels are on the heavy side in both engines, aiding low-speed tractability which the machines are famous for. Though frame geometry has gone back to that of the Sherpa T, the Alpina 350 retains its 348-cc engine displacement, as opposed to the Sherpa's 326-cc unit. The 4 mm-longer stroke in the Alp helps smooth power impulses for trail application and gives about three more horsepower. The Alpina 250 and Sherpa T 250 share the same identical engine unit.

One thing that is a hassle is the fact that fuel and oil have to be mixed by the rider. This is okay on a race bike but is getting to be tiresome in this day and age of metered pumps and oil supply tanks. It's much easier out on the trail if you don't have to worry about mixing up a batch of fuel and whether or not you've got the mixture right. As far as chemistry goes, we used Bel-Ray MC-1 at a ratio of 60:1 and both bikes ran well with little in the way of oil smoke.

Carburetion on the 350 is supplied by Bing and features a handy lever-type enrichening device for cold starting, rather than the old ticklers we came to know so well with early Amals. The latest Amals have the lever choke device too, and the 250 Alp comes with a 26-mm version. Starting is slightly awkward because the transmission has to be in neutral and the start lever is mounted on the left side, meaning that most right-footed riders have to stand on the left side of the motorcycle and kick through with the right foot. Sure, one gets used to it but we still prefer the lever on the right so we can kick while straddling the motorcycle.

Ratios in the five-speed transmissions are the same in both Alpina models and each shifts on the left, as federal law mandates. Old time Bultaco owners who prefer it the other way around can easily switch over to right side shifting with a changeover kit, which puts the brake pedal on the left. Take your pick. Clutch pull is moderate and the gear change lever has to move through a sizable arc to select the next gear. But the ratios are well suited to the engine and the type of riding most Alpina riders will opt for.

Handlebars are on the wide side, unusual for a bike designed for tight confines. Some owners might want to saw an inch or so off the end of each side of the bars, tucking them out of harm's way. Another good idea would be to fit an accessory brand skidplate that will protect the tender chrome-moly frame tubing and expensive outer engine cases.

Bultaco has a nice 60-day parts and labor guarantee that covers the entire motorcycle, unless there's been abuse, misuse or neglect. That's an important factor to consider when purchasing, especially when one realizes how many pure dirt machines carry no warranty whatsoever.

Sure, in some respects the Alpina is an oddball, because it's specialized in a general sort of way. But with the right terrain this could be the right bike. And nobody knows better than us how much fun an Alpina can deliver, except of course, other Alpina aficionados.

FRONT FORKS

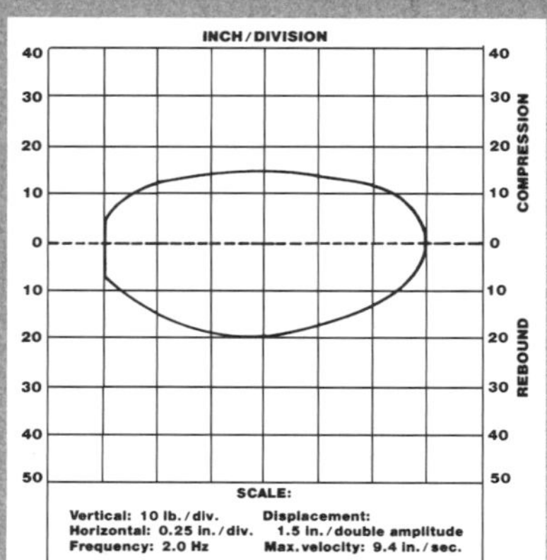

Description: Bultaco Alpina telescopic fork, HD 315 oil.
Fork travel, in.: 6.5
Engagement, in.: 4.5
Spring rate, lb./in.: 18
Compression damping force, lb.: 14
Rebound damping force, lb.: 20
Static seal friction, lb.: 19

Remarks: The stock forks are well suited to the motorcycle, and will be acceptable in most cases. An improvement in front end action is possible, though, by reducing static seal friction. Two seals are used in each of the legs, stacked one on top of the other. Gutting the bottom seals and using them for spacers will cut friction by half. If excessive leakage occurs, a switch to Honda seals should affect a cure.

REAR SHOCKS

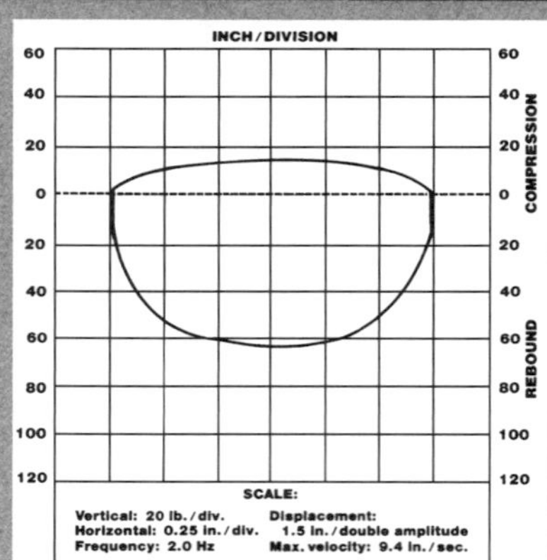

Description: Betor shock, gas pressurized with reservoir
Shock travel, in.: 3.25
Wheel travel, in.: 5.8
Spring rate, lb./in.: 60
Compression damping force, lb.: 14
Rebound damping force, lb.: 64

Remarks: Keeping the Alpina's intended use in mind, the shocks perform satisfactorily. There is, however, a tendency for bottoming under severe compression loads. The cure would be a pair of progressive springs, with an initial rate comparable to the stockers. Handling characteristics would then remain unchanged, and bottoming eliminated.

Tests performed at Number One Products

A little of everything

Frank Melling rides the latest Bultaco Pursang

IT would have been nice to follow last month's interview with Malcolm Davis with an article which reported on his bike in glowing terms. Nice it would have been, accurate it wouldn't, for this latest offering from the Bultaco factory is not the most competitive motorcycle to have come out of Barcelona.

There is nothing wrong with the Mark 10 in that it steers well, is easy to handle, stops just like a good moto-cross machine should and has a motor which produces the right sort of power. The problem lies in the fact that a number of other machines steer better, are even easier to handle and can out-brake and out-perform it to a quite embarrassing extent. What the Mark 10 does achieve is excellent — the trouble is that it doesn't achieve enough.

It is an open secret that the Mark 10 is a stop-gap machine to keep the marque in open class racing until the new 400 comes on the scene and to be perfectly honest, this is what the bike feels like. Three years ago, it would have been a winner but now it is rather dated.

Externally, the Mark 10 looks very similar to the Mark 9. In fact, it doesn't look so dissimilar from the Mark 5, which was the first of the new generation Bultaco racing machines. However, things have got much better and the somewhat unenviable repu-

Said to produce 41bhp at 7500rpm, the Mk10 engine looks very much like Bultaco moto cross engines of the last five years.

tation the Mark 9 earned the factory for unreliability will hopefully be a thing of the past.

Starting from the front, we find one of the best front hubs in moto-cross in the form of Bultaco's own 5½in unit with its unusual sprayed chrome liner. This hub is extremely light, having no iron liner, and is an excellent anchor.

This year, Bultaco have fitted new front forks with 9.6in of travel. They are made for the factory by Betor and compare favourably with most of the current opposition. Like a number of other serious designs, the new forks have dual rate springs giving a sensitive and responsive action over a wide range of conditions.

The rear suspension has still not been sorted out to the extent that it should have been and on Malcolm's bike, Girling gas shocks, specially modified by Reg May, Comerford's competition manager, replace the standard Betor dampers.

This year, the steering head has been strengthened in an attempt to cure Jim Pomeroy's propensity for breaking this part of the bike. However, not too much should be read into this modification since none of Bultaco's other contracted riders ever managed to break a frame at the steering head, so the chances of the average runner achieving this are somewhat remote.

One word of warning which is worth passing on is that the frame will break if the engine is run without the head-steady in place. Malcolm's storeman collected a broken leg through this last season, when the frame broke after a big jump. Needless to say, he had omitted to fit the head-steady.

The motor, whose reliability was subject to question in the Mark 9 version, has now been extensively re-vamped with a view to curing its earlier problems. Worst of these was gearbox mainshaft failure which was due, so the factory diagnosed, to the mainshaft flexing under the load put on it by the drive chain in conditions of extreme suspension movement. The new mainshaft is of massive proportions. The primary drive has been modified to reduce the speed at which the mainshaft revolves and the internal gear ratios have been altered to take this change into account.

Ignition is still by MotoPlat flywheel generator and this contact-breakerless system will no doubt be as good as ever. Like all big Bultacos, the Mark 10 uses twin sparking plugs firing simultaneously.

The cylinder barrel has also come in for some attention with the emphasis being on mid-range power. The charge is now ducted left and right, in addition to the normal centre port so that scavenging is improved.

Technically then, the Bultaco is very much as before, which means very much as the early '70s. Unfortunately, in riding terms, the bike is also of this era which means that the rider is at a distinct disadvantage compared with the best of current machinery.

I did not settle easily into riding the Bul., mainly because it was set-up for Malcolm, who still has the traditional right-hand gear change and left-foot braking. This in itself is indicative of the motor's age for although there is a facility for a left-hand gear change, the bike functions most naturally with the gear lever on the other side — a characteristic which it shares only with CCM. Since I now never have the opportunity to ride a right-hand change bike, the experience was unfamiliar and it took some time to settle in before I could feel sufficiently confident of braking and gear changing to indulge in some fairly hard riding. Malcolm also has the rear brake — a formidable stopper in its own right — adjusted to hyper-sensitivity so that merely showing one's foot to the brake pedal locked the rear wheel solid. Thanks to the ever-efficient skills of Malcolm's mechanic, we de-tuned the brake a little and raised the gear lever so that I didn't knock it into false neutral on every other change and the test proceeded.

The Pursang I rode was Malcolm's own bike and although supplied by the factory, it was virtually a standard Mark 10 screwed together very carefully. In other words, a very representative example of a good 370 Bultaco. This being so, the results were even more disappointing than they would otherwise have been. The biggest fault was the lack of power which strangely was a greater detriment to an ultra-non-competitive rider like me, than an international runner like Davis. At first sight, this might be difficult to understand but in fact, soft usable power is the greatest asset a rider can have in any form of motorcycle sport. For moto-cross, its value cannot be over-estimated since not only does it permit the rider to go quicker, but it also helps him to sort out many of the embarrassing tangles amateurs frequently get into.

TECH SPEC

Engine: piston port, single cylinder, two stroke with petroil lubrication.
Bore and Stroke: 85mm x 64mm. 363.16cc.
Compression ratio: 10:1.
Claimed power: 41bhp at 7,500rpm.
Carburettor: 36mm Amal Mark 2.
Gear Ratios: 1:0.428; 2:0.537; 3:0.670; 4:0.833; 5:1.000.
Wheelbase: 56¼in (1405mm).
Overall length: 83½in (2090mm).
Ground clearance: 13¼in (330mm).
Footrest height: 16in (400mm).
Petrol tank capacity: 2.1galls.
Claimed dry weight: 224lb.

The engine doesn't suffer from gassing up but there is the most noticeable lag in acceleration as the revs build up and the motor drags itself into the power band. Dave suggested revving it harder and things got better when I did this, but not markedly so, since it is easy to go beyond the power peak and get more revs but no more acceleration. Only in the mid-range is the power competitive and it is in this part of the rev range that the bike can be most easily and competitively ridden.

Handling is also less than competitive with the rear end being noticeably inferior to that of the current machinery. Some of the problem lies in the tyres. Our bike was shod with an old pattern 4.60 x 18in Trelleborg, a wonderful tyre when it first appeared but now decidedly outdated when compared with the latest 5.00 and 5.60 section tyres. So a question: "Why fit a 4.60 tyre when there are bigger and better products on the market?", must be asked. The answer is simple. The Bultaco's chain lies so near the rear tyre that any bigger tyre — even a modern 4.60 with large side-wall knobbles — would fetch off the chain.

As a rider, I still like the Bultaco as much as I have ever done. After an hour on the bike my mental computer had been programmed to take into account the fact that the motor pulled best above 2,500rpm and that the rear end would float around more than I would otherwise like and having acknowledged these traits, I was going virtually as fast as I would have done on any moto-cross machine. Riding a Bultaco is always safe. They stop quickly, handle predictably and have no really malevolent vices: just the characteristics which make a top-class enduro machine.

These are also attributes which the novice should seek and it is in this market that Bultaco's best prospects lie. The 370 is cheap (relatively!) to purchase initially, has a good reliability record in its Mark 10 form and spare parts are readily available and modestly priced. At centre level, the Bultaco will be far superior to the riding ability of the competitors and the low running costs could well make up for the fact that it won't win a Grand Prix.

TRIALS REPORT

SHERPA 350

QUITE LITERALLY, the Bultaco needs no introduction — even a die-hard road racing fan knows that the name is synonymous with success in the trials field. Indeed, this factory spearheaded the Spanish invasion of the British trials scene, and has remained as the most popular trials iron ever since. In short, it is the yardstick by which most other machines are judged.

The new Sherpa T 350 is in fact a 326cc single cylinder two stroke. So what? you say, the way it performs makes no difference, but why not call it a 326cc? And it does perform well. From cold, just turn the petrol on, push down the choke lever on the Bing carburetter — no throttle is needed — and give a maximum of three kicks. When warm a single prod is all that is required: the power is there whenever it is needed.

Turn the throttle gently and the power comes in with a smoothness and reliability that is very reassuring even at the most difficult of moments. It will scream up a hill in second or you can make it pull in third: either way the motor will not complain. It just depends on the way you prefer to ride it.

The twistgrip is made of metal rather than fast wearing plastic and has a nice light action, but a strong enough return spring to give a good feel. Gear change is on the left side, and gives a good positive change. You know which gear you have selected while you are positioning to go into the section, and should you need to change once in a section it is very difficult to miss or go through a gear. The lever is very easy to find with your boot as it sticks out a good inch from the timing cover. However, it has no protection at all, so the odd passing rock or tree stump is just as likely to find it as you are, and select another gear or worst still, neutral. If you do find this happening too often there is a conversion kit available to change your gear lever and footbrake from left to right, and this greatly reduces the chances of a knock as the cases then provide protection. The clutch has a nice smooth feel to it, and if you do 'clutch' a lot, then it is not too fierce.

The 'Bull' rides on two Dunlop Universal Tyres — 2.75 x 21in front and 400 x 18in rear: it is stopped by two very adequate 125 x 25mm internal expansion brakes which work just as well in the wet as in the dry. Both front and rear brakes have a very good progressive action and do not grab, but they must be dried out after each trial or the chrome linings of the brake drum will soon rust and flake off, thus causing brake grab and much pain to the wallet as the only cure is to buy a complete new hub.

The suspension to the front and rear feels very good. Two hydraulic Betor dampers keep the rear swinging arm with its 4in of travel under control and nicely on the ground, while the front hydraulic forks with 6½in travel cope with most terrain very well. No problems at all have been experienced in removing or replacing either wheels, and chain adjustment is really simple.

The red plastic Bultaco mudguards do the job, but if bent far enough for long enough will break. Answer to the problem — get your wallet out. The front guard has a mudflap moulded on its rear end and the two thick chrome plated stays will outlast numerous new sets of mudguards. The footrests are very strong and spring loaded with serrated edges — unfortunately they are cup shaped and fill up with mud very quickly as they only have a small slot in the bottom of the cup for the mud to be forced through.

With a ground clearance of just over 12 inches it has no difficulty with rocks and will ride over tree trunks on its not too big, but adequate, bash plate. Thanks to a seat height of only 32½in it is easy for even the shortest of legs to paddle their way out of trouble, but you won't want to sit on the rock-hard seat for too long.

General appearances are good, with slim, clean cut and

Cutting down the steering stops helps in really tight turns.

Left: sturdy stays support the plastic mudguards. Note leading fork spindle.

Right: brake lever is nicely tucked away, and rear wheel spindle is easy to align.

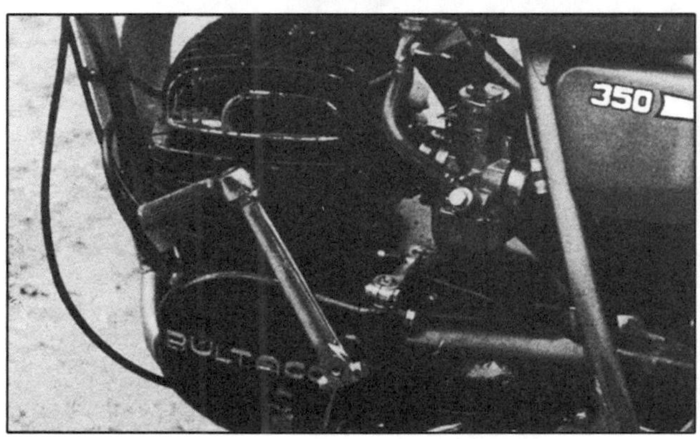

Below left: exhaust pipe follows a tortuous route, but the noise emitted from the final mini box is commendably quiet.

Below right: matt-black paint on the engine does not last long.

Relax Nick! Our intrepid tester demonstrates how easily the front wheel may be lifted — a convenient concrete block being cleared without the sump shield touching down. Mind you, it was close!

very workmanlike lines. Paintwork on the petrol tank is of very good quality with matching bright red side covers and self-coloured plastic mudguards, but the silver paintwork on the frame and the matt black on the engine and silencer is very thin and soon starts to come off, and rust, which if left unchecked can make the bike look secondhand very quickly. All of the aluminium parts which have not been painted have a very good highly polished finish and after washing and drying they come up clean and sparkling. Welding on the frame itself is not in the same class, and though no doubt strong enough, in appearance resembles bird droppings.

Problems? Virtually none, even after having ridden the machine in over eight trials in as many weekends. The exception was front tyre creep, which on one trial actually ripped the valve from the tube. This was cured by screwing two self tapping screws through the wheel rim into the bead of the tyre. I was told when I bought the bike that I might have problems with the petrol tank pouring what looks like mini cornflakes into the carb and filters, but I have had no problems at all. However, as a precaution, I did clean out the whole of the petrol system immediately I bought the bike. I have not had to touch the carb since — except to adjust the slow running and mixture after about fifty miles. However, be warned, as a clubmate did find his

Plenty of torque and grip enables climbs to be tackled with confidence — below is result of a long steep railway embankment.

carb blocked after a few minutes of running.

The only other work I have done is just normal maintenance; oil and adjust the chain, grease the swinging arm and speedo cable and give it a good clean and inspection after each trial.

When you come to grease the swinging arm bushes you will have to change the original straight nipples for right angled ones, unless you have a mini grease gun, as there is only about ¼in clearance between nipples and the engine side covers. Another small mod you can do is to move the speedo head from the original position (on top of a fork stanchion) where it is very prone to be knocked and the bracket broken. This is also very easy to do — you can move it to the centre of the handlebars and have the drive cable coming up the centre of the fork yokes using the original cable and mounting bracket, or you can move it right out of the way down on the front engine mounting.

Another, and I consider the most important mod, is most important modification, is to cut about ¼in off the steering lock stops so that the forks do not *quite* touch the petrol tank on full lock.

As for the riding, I should perhaps first point out that I am not a factory-retained Super Star, but a very average club rider who has been riding trials for two years, and was classified as being of Expert status some six months ago. In short, the sort of person who actually digs his hand into his pocket (or his bank manager's pocket) and buys such machines.

When I rode it for the very first time it was with a great deal of trepidation. I was very apprehensive — and careful — of the power deliver, as it has a reputation of being quite lively, and too powerful for a Novice. However, I very soon became used to it and found the machine taking me up what I had previously thought to be impossible climbs, wet or dry. The more I rode it, the more I learned how to use the flexible power and handling ability, and consequently the more 'cleans' I now get. It really is an incredibly good bike, quite capable of doing most of the work for you.

Besides the good motor and handling ability over all terrain, including the road I might add, the brakes work incredibly well even after being in and out of water during a trial.

What else is there to say about such a well proven trials bike as the Bultaco? You just point it, and if you can ride it, the 'Bull' will take you there. With a wheel base of 51.7in and an overall length of 78.9in it can at first seem a bit tight round the nadgery but with practice and a lot of body lean it soon comes together and gets you round round most things.

NICK JAMES

SPECIFICATIONS

BULTACO SHERPA 350

ENGINE

Type: Single cylinder air cooled all alloy two stroke with steel liner. Caged roller big and little end bearings with pressed-together crankshaft running on ball races. Piston controlled porting.
Capacity: 326.2cc
Bore and Stroke: 83.2 x 60 mm
Compression ratio: 9:1
Carburation: 28 mm Bing T-84 By-pas with mechanical choke, breathing through twin oiled foam filters.
Lubrication: Petrol/oil mixed in ratio of 20:1
Electrics: Femsa flywheel magneto. Contact breaker and coil ignition
Claimed power: 20.8 bhp at 6,500 rpm, maximum torque 20 ft/lbs at 4,000 rpm

TRANSMISSION

Primary: Chain at 2.375:1 reduction
Clutch: Wet multi plate.
Overall gear ratios: 1st 40.7, 31.3, 24.2, 14.8, top 10.7 to 1. Selection by left side, one down four-up foot pedal.
Final drive: Exposed, unlubricated chain. 3.538:1 reduction ratio.

FRAME

All welded steel tube duplex cradle with single down-tube.

SUSPENSION

Front: Telescopic, hydraulic forks with 6.5 in (165 mm) travel.
Rear: pivoted fork with Betor hydraulic dampers providing 4 in (100 mm) travel.

WHEELS

Front: 2.75 x 21 ins. Dunlop Trials tyre on alloy non-valanced rim. Cable operated 125 x 25 mm single leading shoe drum brake.
Rear: 4.00 x 18 in. Dunlop Trials tyre mounted on alloy non-valanced rim. Rod operated 125 x 25 mm single leading shoe drum brake.

EQUIPMENT

1.1 gallon steel fuel tank, flexible plastic mudguards, alloy handlebars,folding foot rests, alloy levers, alloy sump shield. Lighting set available. Right hand gear change conversion set available.

DIMENSIONS

Weight (claimed dry): 203.9 lbs (92.47 kg).
Wheelbase: 51.7 in (1313 mm)
Ground clearance: 12.2 in (310 mm)
Seat height: 32.5 in (825 mm)

CUSTOMER INFORMATION

Importers: Bultaco (UK), Oxford House, Portsmouth Road, Thames Ditton, Surrey. Tel: 01-398 5531.
Colours available: Red/Silver
Price: £840 including VAT. Delivery charge £14.00 extra.

SPARE PARTS (Inc. VAT)

Piston (complete with rings, pin, clips etc)	£24.05
Clutch plates (6)	£14.27
Complete exhaust system	£39.08
Auxillary silencer only	£16.42
Contact breaker points	£2.43
Gear change pedal	£5.01
Rear brake pedal — alloy	£8.02
Rear brake pedal — steel	£4.13
Mudguards — front	£5.29
Mudguards — rear	£7.34
Footrests (each)	£1.12
Brake shoes (pair)	£3.19

BULTACO ALPINA 350

One of the Best Do-It-Yourself Woods Bike Kits

by Henry N. Manney III

■ At the recent automobile GP in Long Beach I happened to run across the celebrated John Surtees, who used to be into bikes before he was into racing cars and is, I think, the only World Champion around in each speciality. Meeting for the first time in nine years, we fell upon each other's necks and almost the first question was, What bikes have you got now? Well there was this and there was that and then he had a couple of Bultacos. I should have known. Running a Formula 1 car team is a pretty masochistic exercise and Bultaco owners have a masochistic streak a mile wide. Not for us the carefree abandon of an empty toolbox on sallying into the wilds; every Bul owner worth his salt carries at least a heavy-duty plug wrench and spare plug while we confidently expected the latest Alpina to have a cosy fitted box for jeweler's screwdriver, flywheel puller, blowtorch, 5 lb. bronze hammer, vernier tdc indicator and two or three sets of ignition parts, not to mention a few extra jets. Actually it isn't really that bad as Senor F.X. Bulto *is* a nice man besides being an enthusiast; you feel that Bultacos are built by human beings instead of by computer-fed automatic machinery. Furthermore he does listen to remarks from the outside world which means that this latest Alpina is a considerable improvement on earlier ones even if it took a while, carving our way through a wall of human flesh, to dig the pearl out.

Alpinas as you probably know are sort of a hybrid between enduros like the Frontera and the Sherpa trials mount. As such they enjoy a relatively limited market

Photos by Ron Hussey

Exhaust pipe is bulky and awkward but it is also routed away from the rider and from harm.

The Alpina is a unique mixture of trialer and enduro-legal woods bike.

Alpina engine has been tested and refined for years. Provided it's jetted and timed correctly, it will pull from 25 rpm on up.

but would do better if more customers realized that what they really should be riding on their woodsy potters is an Alpina and not a fire-breathing motocrosser, much less a 90 percent street bike decked out as an Enduro. Basically the 350 Alpina is a 350 Sherpa Trials with very few detail changes (triple clamps, bars, lights, knobbies, slightly bigger engine capacity) and as such is not supposed to go flat out across the desert or, for that matter, flat out anywhere. The engine is ported for low-speed flexibility, the quiet Sherpa pipe ditto, and thus pays rewards in meandering gently through the woods communing with nature, although for connecting road sections you can get it up to about 60. Who wants to run into trees at 60?

Our Alpina arrived in a box (promptly appropriated by Randy Riggs to store old dollies in) and provided me with a nice afternoon's entertainment putting it all together. Once done, the Bul appeared a very handsome bike indeed and unlike the last lot we had (see tests April, 1977) fired up second kick without an orgy of cursing, bump starting and incipient cardiac arrest >

although clearly jetted so rich that it must have been set up for the shores of the Dead Sea. A close survey revealed that there had indeed been some changes from last year, i.e. rim locks instead of those blasted rim screws to keep the tires on, a return to the hard and narrow Sherpa seat, a sexy black paint job on the whole engine, plastic fenders instead of the lovely but impractical alloy ones, an extremely lightweight headlamp, giant taillight mounted on a rubber strap (our desert experts wanted these) and a slightly different paint job. No primary kick, no oil injection, no overhead cams but then it wouldn't be a Bultaco, would it?

As always with Buls you wonder how the hell a factory making off-road machinery could pull some of the odd tricks found on the Alpina. For example Bultaco boasts that they have enlarged the fuel capacity but even so, 2.24 gallons isn't nearly enough to enjoy a day's riding without returning to the friendly leaking gas can. Especially here in the West, a rider can easily make a hundred mile detour around a mountain before elevenses and be, say, 75 miles in a straight line away from his premix. Buddy pegs, especially cheapo buddy pegs, look very odd at the bitter end of the swing arm. The tichy little toolbox has most of its access blocked by the lid tabs inside. A fork lock on a bike with no key. Junky cable adjusters which, if the cable breaks, become lost in the sand. A headlight rim which unscrews itself, spilling the sealed beam out like a hanged criminal. A self-shorting kill button, *and,* under the heading of service, fork caps unremovable without shipping the handlebars. A gearbox filler difficult to unscrew without ditto the exhaust pipe. A soft plastic clutch adjusting plug. Having to pull the rear brake pedal, incorporating a double-jointed split pin, to remove the

primary side case. Having to screw the bar clamps down really tight to prevent El Foldo when you may not want them that tight to avoid bending in a crash. Having to use an open end for the gb drain instead of a socket because the bash plate hole isn't right. And not everybody liked the bulgy side panels.

BULTACO ALPINA 350

SPECIFICATIONS	
List price	$1290
Suspension, front	telescopic fork
Suspension, rear	swing arm
Tire, front	2.75-21
Tire, rear	4.00-18
Engine	two-stroke Single
Bore x stroke	83.2 x 64mm
Piston displacement	348cc
Compression ratio	8.5:1
Claimed power	21.6 bhp @ 6500 rpm
Claimed torque	20.1 ft. lb @ 4500 rpm
Carburetion	28mm Bing
Ignition	Motoplat magneto
Lubrication system	premix
Oil capacity (transmission)	1.3 pt.
Fuel capacity	2.2 gal.
Recommended fuel	premium
Starting system	non-primary kick
Air filtration	oiled foam

POWER TRANSMISSION	
Clutch	multi-disc, wet
Primary drive	3/8 x 7/32-in. chain
Final drive	#520 chain
Gear Ratios,	overall:1
5th	7.67
4th	9.35
3rd	12.28
2nd	17.36
1st	26.74

DIMENSIONS	
Wheelbase	51.5 in.
Seat height	33.2 in.
Seat width	6.4 in.
Handlebar width	34.5 in.
Footpeg height	13.6 in.
Ground clearance	11.8 in.
Front fork rake angle	26 deg.
Trail	na
Curb weight (w/half-tank fuel)	239 lb
Weight bias, front/rear, percent	45/55

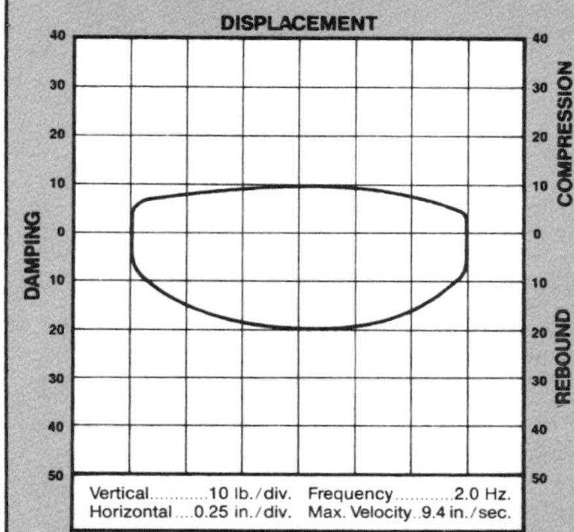

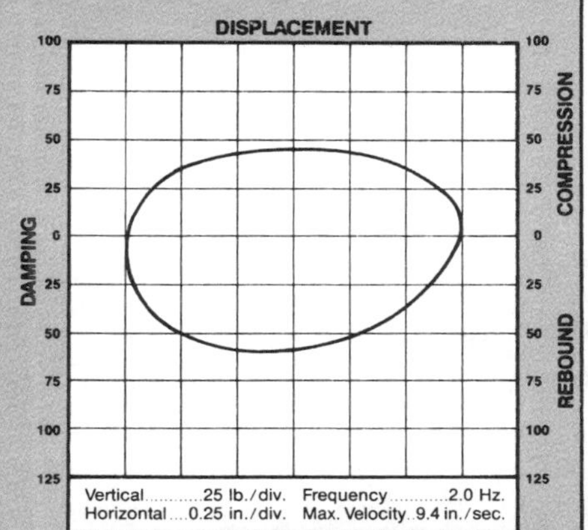

FRONT FORKS	
Betor leading-axle fork	
Fork travel	6.7 in.
Engagement	4.3 in.
Spring rate	14 lb./in.
Compression damping force	10 lb.
Rebound damping force	20 lb.
Static seal friction	24 lb.
Stanchion tube diameter	35 mm

The Alpina's forks are potentially good but excessive seal friction and heavy oil make them far too stiff. Gut the top seal in each leg, or swap the stockers for a set (one per leg) of accessory or Honda seals. Use 10 wt. oil to provide proper damping rates and fork action.

REAR SHOCKS	
Betor shock	
Shock travel	3.8 in.
Wheel travel	6.6 in.
Spring rate	68/110 lb./in.
Compression damping force	44 lb.
Rebound damping force	60 lb.

The Bul's rear end is too stiff for most riding conditions. The dual-rate springs are fine for the bike's weight, but the shocks' excessive compression damping prevents proper suspension action. A set of aftermarket shocks would be worthwhile.

Tests performed at Number 1 Products

On the credit side, the finish is marvelous, the neat plastic headlight adjusts up and down, the non-shock speedo mounting is clever, the pegs are the best in the business, two weeks' running richer than Croesus didn't soot the plug, the handy grab rail is nice, there is a proper little tensioner for the primary chain, vent tubes sprout from both side cases, cotter keys and/or stop nuts abound, the engine was timed properly and the points haven't lost their gap yet (and a good thing too as now there is a new flywheel ign with only one tiny window in it; the points gap, however, is cast on) and the fuel filler cap doesn't have one of those tubes hanging from it, yet one more thing to move. And of course, it is a Bultaco which means a lot when you start down one of those gullies you shunna started down.

We first went up into the mountains to take photos and in spite of our dropping the needle in the Type 84 Bing as far as it would go, the poor 350 was still rich as all buggery. Furthermore, as delivered the suspension was harder than a Harley Chopper's which meant that in spite of letting air out of the Pirellis until they "pooched" at rest, the bike was still a handful on anything washboard, sandy, ridged, rocky or gravelly as either the front or back wheel would hop violently sideways at the odd pebble. I don't know what the factory was thinking about. The consultant services of Ron "Crasher" Griewe were called into play and we drained what looked like 90 wt. from the forks, removed two of those little spacers from each leg, and filled up with 5 wt. Bel-Ray (20 is specified) so as to begin with Too Soft. Too Soft is right as the fender would hit the frame on braking, let alone whoops, and there was the interesting display known as Pogodie Pogodie Pogodie Thud. One spacer added back in was still too soft really so both were restored with the result that the forks are lovely and responsive for slow trail riding or rock picking but will bottom given sufficient reason. Probably 10 wt. is the answer or even 20 with one spacer out, all depending on what the Empirical Research Institute has to say. Like Caesar's wife, dirt bike suspension is all things to all men and each rider has to fiddle around a bit to suit his own needs, not just accept what the kid at the parts department wants to sell that week.

Likewise the rear suspension was punishingly hard, perhaps to suit the buddy pegs, but surprisingly enough became a lot more efficient when the front was softened up. Perhaps because of better shocks, the rear wheel control on the 350 (as opposed to the 250 Alpina I had) is pretty good and young Oley the trials rider pronounced it "just right" for trials work even if a bit hard for trail riding, curiously enough. Perhaps when the seals wear in.

Carburetion was another problem as we dropped the needle as far as it would go without any results, really, except losing that diabolical plastic bucket that lives in the slide. Do you know how strong that spring is? and how far a plastic gizzie will go in a cluttered garage? Mike Kane at Va. Beach however had sent along spare jets and the next size leaner needle jet with the needle in the next place from top put it in the ballpark for around here. At one point the throttle cable mysteriously shrunk about two inches, causing a lot of bad language and a general disassembly; the trouble eventually proved to be the plastic tit guiding the slide's developing a burr on its leading edge, hanging up said slide about halfway down. The Bing is very sensitive to needle position, one notch being the difference between no power above 1500 rpm and well-nigh perfect carburetion and all I can say is that the Bing is a powerful improvement on the 250's Amal.

All the battle is worthwhile when the Bul is finally right (or close to it) as the 350 engine is smooth, responsive, and provides enough unfussy low speed grunt in third even, picking its precarious way up through a trough full of baseballs. The remarkable thing about Bultacos is that they go precisely where they are pointed and do so without being deflected too much by natural hazards. Naturally the rather hard-mix Pirelli knobbies don't give anything like trials feeling but a couple of semi-pros who rode the Alpina felt that it needed only a wider steering lock plus removal of some of the weight (speedo, light, an alloy tank perhaps) off the front end to be as near as dammit. CW contributor Chuck Johnston also had a bash, doing wheelies all over the place commenting that this was a *real* 350 Bultaco and climbed a couple of impossible hills to prove it, mentioning only on coming down that it was time Buls built a better rear brake. Well it *does* provide gentle retardation even if tending to go away after water splashes, the front one anyway working very well even for occasional road use. (The Alpina ain't legal here without battery, horn, and mirror, and I think the lights are extra in the U.S.) The 5-speed gearbox shifts faultlessly from the left (thank you Uncle Sam) but can be modded to shift from the right for traditionalists; its normal knife-through-butter operation was marred somewhat by a sticky clutch which made getting neutral at rest a dodgy business. We have to look into it. Every Bul I have owned had a super clutch.

Well. Bultaco is a state of mind and my mind is very pleased with the 350 Alpina at the moment, thank you, as it runs sweeter than any two-stroke I have ridden since my old Mk 4 Matador. Power is abundant for anything besides half-mile as the torque curve goes straight up, and the Sherpas didn't get their reputation for good handling from advertisements. Isn't it comforting when the bike's potential is greater than the rider's? Who's this guy Sammy Miller? ◘

Bultaco should hang a "for-men-only" sign on the 325 Sherpa T. There is so much power on hand that it can get you in more trouble than it will get you out of, if you're not careful. For experienced riders it's an absolute delight. Weight of 214 is same as Ossa, 34 pounds more than the 123 Montesa.

BULTACO'S BEST YET

Bultaco 325 Sherpa T

Four world championships, a second and a third spot say an awful lot about a bike. In particular what can be done to improve it. When you're at the top of any league, there will always be someone or something threatening to push you back down the ladder. Bultaco have always been conscious of this fact of life, probably more so than any other manufacturer. With world champions Vesterinen Schreiber and Lampkin who between them have 'cleaned up' the majority of this years world championship rounds; the policy remains the same with the continuing search for perfection.

The new '79 Bultaco 325 with its dazzling blue finish makes it visually agreeable to the eye and therefore instantaneously more recognisable than the previous model. Before I get down to the 'nitty gritty of riding the bike, the improvements must be mentioned as they make a substantial difference to the handling of this 'works type bike'. I say this because of the similarity between the production bikes and the bikes used by: Vesty, Schreiber and Lampkin, whilst campaigning around the world in 1978. It must be pointed out, that 'big Martin' has a 350 motor in his bike, but chassis-wise, they are all the same.

First let's mention the changes made in the engine; basically it's the same 325cc 2-stroke, breathing through 28mm Bing carb with wide ratio 5-speed gearbox, now the Bultaco has always been renowned for its low-down and mid-range pulling qualities or torque; but top end power, or its ability to rev out has brought criticism on earlier models, this has been overcome by: re-working the ports, redesigning the air filter and enlarging the exhaust pipe between the two silencers. With these mods the motor retains all of its torquey benefits with the added bonus of being able to breath easier at the top end of the rev range; and very handy it is too.

The real changes, however, are in the chassis. With world championship trials getting tougher by the event, trick type sections are much in evidence so that frame designs have to follow suit. To keep up with this, Bultaco have drastically changed the new bike's frame layout. Firstly the engine has been lifted 8mm in the frame and the tubes that run under the engine have been removed and replaced by a plate that runs the whole length of the engine's base. These things together give a phenominal 13 inches of ground clearance and are much neater in appearance.

The next change made, was at the swinging arm spindle, which has been re-located a half of

New steering geometry gives quicker response but front wheel lift is easily controlled by body weight, in fact if you look closely you can see 'Taco is capable of climbing trees.

Evidence of the bike's pulling power – the above climb has been taken in third gear. Right: Harris up to his usual 'off the wall' antics.

an inch higher in the frame, this has been done to increase the rear suspension movement. The swingarm itself has been beefed up with gussetts welded on top of the swingarm tubes either side of the rear suspension lugs.

Now with this extra movement on the rear end, two chain slipper pads are necessary, one on the swingarm at the point where its spindle runs through and another on the base plate that runs underneath the engine; all very neat and tidy. A nice point is the new flywheel ignition cover plate, this follows the contours of the flywheel. But the ultimate improvement must be in the steering head angle which has been reduced, giving the rider very responsive steering with virtually no caster action, (making the bike extra responsive in the trickiest of situations). With this taken into account, the handlebars have been reduced in overall width and have been braced to match. With such light steering, who needs all that leverage that comes with wide handlebars?

The wheels are the same as on the previous model, the much tried and tested ones with alloy rims, hubs and chrome lining on the drums; light but surprisingly strong.

The front forks on 'Bully's' have always been the envy of other manufacturers and so there is no change here, as with the very light oil in them they react perfectly in gripless mud or charging through the rockiest ravines. The rear units are the same formula as the '78 models and cope extremely well with extra rear movement.

The tyres are 'supergrip' Pirelli as used now by most of the manufacturers running on 7psi in the front and 4psi in the rear. (They will uncannily find grip where none exists). There was no creep on either rim during testing, although I would not be adverse to drilling the rim and putting in a self tapping screw just to make sure that the tyre stays in place.

Starting the '78 bike was a bit of a problem if you were riding with a left hand gear change (Comerfords do a conversion for the right hand side). When kicking the bike over, you invariably ended up with a dead engine and the bike in gear. Not so anymore, the '79 bike has been modified to alleviate this annoying trait.

Now it's getting your boots and hat on time, and a great day it was. A bright, clear blue sky, with autumn leaves falling in the very cold

BELL

wind; perfect for riding.

The bike that was being used for the test was Colin Boniface's own; the young super star had given the bike a taste of blood the previous week with an outright win in the "Charlie Wake Trial", a south-eastern centre championship round.

The Bully fired up second kick on full choke and it will take 5 to 10 minutes to get the motor and suspension warm. Peter Hudson, Bultaco's sales manager, undertakes this task whilst I tell you about the section and the type of going. Primarily the ground is hard and sandy with steep climbs including rock and root outcrops, also there are foot-rest-high gullies climbing up through the latter, making riding tricky.

The first section that I tried was a very steep drop with rock and roots at the bottom. You have to turn left immediately and come back on yourself riding an earth wall with the front wheel to clean the section. Dropping over the edge of this section with the Bultaco as I normally do, I pull in the clutch slightly – holding the front brake and locking the rear wheel, then with a slight turn of the bars to the left, the whole bike is sliding gracefully sideways into the turn; so that at the apex of the turn the bike has straightened out so much, that all you have to do is: de-clutch release the brakes and squeeze the throttle, not too much, so that the wheel steps out, but just enough to make the tyre grip and off you go, feet up.

On this particular day, the section was very slippery and I could only make it clean in one attempt out of three, so I had another think about it and decided to attempt the section without using the clutch, just relying on the torque. When going over the bank instead of pulling in the clutch, I let the motor run on, holding it slightly with the rear brake and just a little on the front. At the apex of the turn, I was able to take a different line when the brakes were released; round it came smooth, clean and easy every time. I was beginning to feel very confident now, riding in, out and over everything.

Next came hill climbing; starting on the flat and rising gently, faster and faster, steeper and steeper, we go, flat out in 3rd gear. The last six feet are sheer; shutting the throttle just as we hit vertical and for a split second in time up we go, man and machine – flying, an exhilarating sensation! The Bultaco can do this all day,

BULTACO 325

Engine:	2-stroke single
Bore & Stroke:	83.2x60.0mm
Capacity:	326cc
Comp Ratio:	N/A
Carburation:	28mm Bing
BHP @ RPM:	20.8@6500
Primary Drive:	Chain
Clutch:	Wet multi-plate
Gearbox:	5-speed
Electrics:	6v flywheel magneto
DIMENSIONS	
Wheelbase:	51.7ins
Seat Height:	35.8ins
Ground Clrnce:	13ins
Weight:	N/A
Fuel Capacity:	N/A
Tyres (F):	2.75 x 21
(R):	4.00 x 18
Brakes (F):	Drum
(R):	Drum
Price:	£979
Importers:	Comerfords Ltd., Portmartin Rd Thames Ditton Surrey

it will go up everything possible and a few things which are impossible too!

On an adverse camber you can follow a line, but if you deviate either side by even an inch or so, it means instant stop. Body lean gives you instant correction. So rapid is the steering that virtually just a thought puts you back on line. Now don't get me wrong, but this bike is an experts machine – hit the gas too hard and you'll be in trouble! spinning to a halt, or it will grip so hard that you'll be thrown.

It is so sophisticated that it would probably take you a couple of months to get to grips with it, especially the spin and grip techniques that are used in very muddy conditions.

On previous models, the rear end was a bit of a problem, in particular mud clogging the swinging arm and exhaust box; unfortunately I

was unable to test the bike in conditions that normally give this type of trouble which is more prevalent in the Kent area than any other.

I noticed as I was parking the bike the stand had been cranked in towards the frame to prevent it catching on any obstacles, a great improvement.

To sum up – the best bike from Bultaco so far. I would think that this bike will take a production run for the next two or three years, without change; and with tyres coming up for scrutiny from world championship contenders with wet and dry compounds being used – but that's another story.

At £979.59 (plus £16.50 delivery charges) this represents outstanding value, falling into the super trials bracket along with the Suzuki and Montesa.

BULTACO SHERPA T350

Sammy Miller's 1964 Revolution Just Gets Better Every Year

■ Trials bikes would look very different today had not Francisco Xavier Bulto in 1964 invited Sammy Miller to spend a few days at the Bultaco factory in Barcelona. The few days stretched to 12 and at the end of his stay, Sammy rolled the prototype Sherpa T out in front of the factory engineers and Señor Bulto. The concept was rejected at first but after Sammy proceeded to clean sections his factory Ariel couldn't even attempt, the bike was adopted. Sammy brought the bike back to England for its first competitive trial and you can imagine the furor caused when he arrived with the Bultaco and won his very first outing on the bike. Sammy was already a legend in the trials field and his Ariel, GOV 132, was considered the ultimate trials machine—the little Bultaco just couldn't have been more different.

The bike was essentially a modified 200cc ISDT model that Sammy enlarged to 250cc. He also made some gearbox changes, changed the exhaust system and added an extra flywheel to improve low end smoothness and, as Sammy is prone to do, experimented with carburetion. The changes by today's standards of modification were quite minor but the machine ended up dramatically changing the shape of trials machinery. During the first year Sammy rode the new Bultaco, he won the British Solo Championship and the Scottish Six Day as well. It was the first time a two-stroke had won the event and certainly the first time a bike built in a foreign country had won.

Bultaco's only lasting competition since then has come from the other two Spanish trials bikes, Montesa and Ossa. The Japanese giants were in the trials scene for a while with Honda and Yamaha putting forth some major efforts but when the great American trials boom failed to materialize, the Japanese phased out of production and as we write this, only Bultacos and Montesas are available to riders on the U.S. west coast.

We'd been trying to obtain a new Sherpa T for testing for some time but to no avail so when the chance came up to test a privately owned bike we quickly accepted.

We hadn't tested a Sherpa T for three years so we read over our last test and were amused to see we had written that we didn't believe the bike could be much improved. The Sherpa T was then, as it is now, a world class trials mount that can be taken out of the crate and ridden in international competition events.

The Sherpa T lost its fiberglass gas tank last year and we found the new matt finish soft plastic tank less than appealing. Sherpa Ts were always a very pretty motorcycle but we must admit owners did worry about damaging their beautiful glass tanks. Well, gone is the worry about the tank but gone also is some of the sparkle.

The new tank holds only one gallon which is more than enough for the loops laid out in most trials events. It can be removed easily by unbolting the two 6mm seat bolts and one rubber-grommeted bolt running through the tank just behind the steering head. Gone forever also is the crossover tube once used on the trials tanks. The older Sherpa tanks had to be removed to service the air cleaner and that always meant gasoline running down your arm and a mess on the shop floor.

The air cleaner is now reached by removing a right side cover and pulling off a rubber band that holds the cover to the airbox. The rubber band routine is not one of Bultaco's better ideas for 1979 because every new Sherpa T seems to have the air box cover duct-taped in place. The test bike was one of the first 25 in the country

Plastic tank is fall-proof, but lacks the looks of the older fiberglass tank. Righthand shift lever is set high, as some trials riders prefer.

and this may already have been remedied. To Señor Bulto's credit, he doesn't wait for the next model year to change items that need attention and each shipment of Bultacos leaving Barcelona usually sports some small improvements.

The only fiberglass still used on the bike are the panels on each side of the airbox and on our test bike both panels were already showing signs of cracks. Perhaps these will be replaced with the new soft plastic. We complained about aluminum fenders on the Sherpa T back in 1977 and the new bike uses a soft plastic fender front and rear that we tested to the limit in a very muddy Saddleback Park—so muddy in fact that the park was closed to the public. After only a few sections the bike was very muddy but we remained unusually clean for such conditions. Throughout the day, even when we were unable to see the wheels, the bike never became clogged with mud to the extent we had to stop and clear away the muck. Someone at Bultaco has already done mud testing.

The lightweight aluminum rims come mounted with the new Pirelli MT13 trials tires that are becoming quite popular with the plonker set. The MT13 is a serious trials tire, built and designed to meet FIM rules and not meant (or certified) for road use. The MT13 (like the Dunlop Trials Champion, which was designed by the same man) has two plies, flexible sidewalls and is supposed to extend its side knobs when the center of the tread is depressed. Great for off-camber climbs. We can't speak for the Dunlop Champion but the Pirellis we tested certainly do the job in the mud. Our only real problem was making the transition from mud to rock where the rear tire would slip at first but then quickly clean itself and grab traction. We ran them at the normal trials tire pressure of 5 psi in the front and 4 psi in the rear.

Walking up to a 1979 Sherpa T for the first time the thing that is likely to catch your eye besides the new blue color scheme is the removal of the frame tubes under the engine. The frame down tube now ends just in front of the engine cases and is connected to a heavy duty aluminum plate that functions both as a skid plate and frame member. Six flathead allen screws hold the plate in place and the whole arrangement seems quite like the Sammy Miller HighBoy frame that has been available as an aftermarket item for

several years. Can Sammy still be influencing Barcelona?

Whatever its origin, the aluminum plate saves weight and still manages to protect the underside of the engine. Some riders may still want to add further protection for the side cases but all in all we like the new layout.

The frame itself is chrome-moly and decals on each major frame member remind you of that fact. The swing arm is cleanly gusseted. Non-rebuildable Betor shocks are mounted in a manner that allows 100mm (about 4 in.) of travel. The units can be adjusted for preload and we thought the valving and spring rate ideal for trials work. We know a few aftermarket shocks are available for the bike but even the experts seem to favor the Betor unit and we understand why.

The front suspension units are Betor as well and according to Bultaco, have 6.5 in. of travel. This is either a misprint in the Bultaco literature or our test bike had an improved model fork for the wipers showed more travel than this and we never bottomed the forks during testing. Our test riders weighed 165 and 185 lb. respectively and we used the SAE 20 weight oil that Bultaco recommends. No signs of seal leakage were evident and the original seals were still in the bike.

Although Bultaco builds a 250cc version available in Europe, only the 350 engine is sold in the U.S. The engine uses an uncorrected compression ratio of 9.1:1, up slightly from last year, and it remains a conventional piston-ported two-stroke which Bultaco claims produces 20.8 bhp at 6500 rpm. Maximum torque of 20 ft. lb. is reached at 4000. Numbers like these don't always relate to actual riding but rumor is that the 350 motor will pull cleanly from 300 rpm up to top revs and we tend to believe this after spending time on the bike.

The engine modifications for 1979 consist of slightly changed intake porting in order to further improve response from a closed throttle. The carburetion on the 350 engine is provided by Bing and even this unit has been modified somewhat for 1979 to improve throttle response. A cold start lever is used on the 28mm carburetor and we only had to use it once at the start of the day.

Which brings us to starting Bultacos—they start all right but the kicking may be a bit awkward at first. Some riders will step off the bike and use the left side mounted lever with their right foot. The other technique lets the rider stand astraddle the bike and use his left foot. We found the latter method rather clumsy and we had to slip the clutch a bit to get the kick lever to travel part way down its travel. Bultacos have never featured a primary starting arrangement and that means finding neutral each time the engine is killed. Although the kick start lever has been redesigned slightly, our test bike didn't feel any different than Bultacos we have been kicking for 15 or

Sturdy 350 engine has been mildly updated, still works well for trials, and kick lever is still awkward.

more years. Still difficult for medium and tall riders and a genuine problem for the grunt leggers.

The magneto is of conventional design featuring points and condensor with the Femsa-made unit allowing both point and stator plate adjustment. A newly designed side cover over the magneto has a vent tube that can now be removed with a wing nut attachment and the fitting itself has been beefed up.

The motor is as quiet as we have come to expect of the Sherpa T and new rubber blocks between the cylinder fins contribute to this. The exhaust system is effective and does include a spark arrester in the second of its two mufflers. An O-ring has been added to the connection between the two mufflers and we couldn't spot a drop of oil leaking from around the cylinder connection or between the mufflers.

Our test bike arrived in the U.S. as a left side shifter but it was quickly changed over to the right side. Even a quick glance tells you the bike was designed to be shifted from the right because a very strange shaped shift lever must be used to clear the final drive chain if the bike is shifted from the left. Changing the shift lever allows it to be installed at the 45° angle that many Sherpa owners prefer. The lever is much less susceptible to rock damage and the brake lever likewise fits better on the opposite side. A cable arrangement is necessary if the brake lever is on the right and moving it to the left allows a direct pull to the rear brake.

Cowtrailers might leave the levers in the stock position but then we have the feeling most Sherpa Ts are used for competition and for good reason: it wouldn't be all that comfortable to spend the afternoon on that thin and narrow seat.

Whatever your reasons for buying the Sherpa T, the machine is built for competition and more than one world class rider has won major events on a box stock Sherpa T. Arriving late and not getting the bike out of customs can be a reason for this trauma and usually a local dealer will knock open a crate and do a quick set-up. The rider might change the handlebars and do a bit of jetting but it's nice to know the stock bikes are capable of winning.

As mentioned earlier, we tested our Sherpa T in very muddy conditions and what would be a dreary day to any but a trials rider, we found most welcome. Give a trials rider muddy gullies, grey skies, and green hills and you've sent him part way to Scotland.

We've ridden Sherpa Ts for years but stepping onto the '79 model was still a pleasure. The incredible balance is still there and the feeling of lightness (207 lb.) is not an illusion. Readers must get tired of hearing about how precisely certain motorcycles steer but perhaps the Sherpa T has to be experienced to be fully appreciated because it will go exactly where you point it and unless deflected by a good sized rock, is quite easy to steer.

In all our testing the bike only did a reverse lock number once and that only required a minor tug to set it straight again. We used low gear once during the entire test with most of our sections being attempted in second and third gears. We remember a factory rider once saying that if in doubt about which gear to use in a section, you should pick the higher of two and then shift up one. The point is that experts will almost always use a higher gear than novices might think possible.

We understand the final drive ratio in the new Sherpa T is a bit higher to allow for rapid transit between sections but frankly, we didn't get much of a chance to use it in our test conditions. Non-trials riders are often surprised at the speeds trialers ride between sections and at one national event we saw a demonstration not soon to be forgotten. It was late in the day and the late penalty time was approaching.

BULTACO SHERPA T350

SPECIFICATIONS

List price	$1810
Fork travel	6.5 in.
Fork stanchion tube diameter	35mm
Real wheel travel	5.5 in.
Tire, front	2.75-21 Pirelli
Tire, rear	4.00-18 Pirelli
Engine	two-stroke Single
Bore x stroke	83.2 x 60mm
Piston displacement	326.2cc
Compression ratio	9:1
Claimed power	20.8 bhp @ 6500
Claimed torque	19.9 ft. lb. @ 4000
Carburetion	28mm Bing
Ignition	battery/points
Primary drive	#530 chain
Gear ratios, overall:1	
5th	9.9
4th	16.2
3rd	26.3
2nd	34.5
1st	44.4
Oil capacity	600cc
Fuel capacity	1.4 gal.
Fuel tank material	plastic
Swing arm material	chrome-moly
Starter	kick
Air filtration	foam
Frame material	chrome-moly

DIMENSIONS

Wheelbase	52.1 in.
Seat height	31 in.
Seat width	8 in.
Seat length	14 in.
Seat front to steering stem center	19.5 in.
Handlebar width	32 in.
Footpeg height	15 in.
Footpeg to seat top	17 in.
Footpeg to shift lever center	6.5 in.
Footpeg to brake pedal center	5 in.
Swing arm length	16.25 in.
Swing arm pivot to drive sprocket center	6.25 in.
Gas tank filler hole size	1.75 in.
Ground clearance	13 in.
Fork rake angle	27°
Trail	na
Test weight w/half tank fuel	207 lb.
Weight bias, front/rear percent	45.5/54.5

FEATURES

Forks adjustable with air?	no
Rear shock damping adjustable?	no
Rear shocks rebuildable?	no
Provision to check transmission oil level?	no
Does owners manual show how to disassemble complete engine?	no
Does pipe burn rider?	no
Brake pedal height adjustable?	no

Rather than incur late points, Leavitt, Schreiber, and Whaley were racing elbow to elbow between sections over some very difficult terrain. Even local motocrossers stood slack jawed.

With a head angle of 27°, one would expect a bike to be twitchy at anything but a crawl but the Sherpa T could be trusted to a beginner without much worry. We wouldn't want him to ride the bike at desert racing speeds very long but it is quite predictable.

We found no fault whatsoever in the motor, transmission and handling of the new Sherpa T but as usual we weren't all that impressed with the rear brake during our damp test conditions. This won't come as a surprise to Sherpa T owners and we are beginning to wonder why Bultaco doesn't build a better rear brake. For all the shortcomings of the Japanese trialers, they did have back brakes that worked in the wet—Señor Bulto, Que Paso?

On earlier Sherpa Ts the handlebars were one of the first things changed but we think the current Vesterinen bend will be left on many of the bikes. Although on the low side, the new bars suited both test riders.

We liked the 1979 Sherpa T 350 and as usual we find ourselves wondering if the 1980 bike can be much improved. With much of the competition out of production, Señor Bulto could sit back and let the bike stand as is for a few years. We have the feeling that the Sherpa T and trials riding have a special interest to Francisco Xavier Bulto and he'll not likely be content with the current model. But even the Patron will admit it's a hard act to follow.